S0-BPO-517

A HISTORICAL CATALOGUE OF SCIENTIFIC PERIODICALS, 1665–1900

GARLAND REFERENCE LIBRARY
OF THE HUMANITIES
(VOL. 583)

By the Same Author

A Historical Catalogue of Scientists and Scientific Books: From the Earliest Times to the Close of the Nineteenth Century. Garland. 1984.

A HISTORICAL CATALOGUE OF SCIENTIFIC PERIODICALS, 1665–1900

With a Survey of Their Development

Robert Mortimer Gascoigne

GARLAND PUBLISHING, INC. • NEW YORK & LONDON
1985

Library of Congress Cataloging in Publication Data

Gascoigne, Robert Mortimer, 1918–
A historical catalogue of scientific periodicals, 1665–1900.

(Garland reference library of the humanities ; vol. 583)
Companion vol. to: Historical catalogue of scientists and scientific books. 1984.
Includes index.
1. Science—Periodicals—Bibliography—Catalogs.
2. Scientific literature—History—Handbooks, manuals, etc. 3. Science—History—Handbooks, manuals, etc.
I. Title. II. Series: Garland reference library of the humanities ; v. 583.
Z7403.G3 1985 [Q158.5] 016.505 84-48863
ISBN 0-8240-8752-6 (alk. paper)

Printed on acid-free, 250-year-life paper
Manufactured in the United States of America

CONTENTS

Introduction vii

List of Tables xi

PART 1. THE CATALOGUE

Science in General 4

Mathematics 31

Astronomy 35

Appendix: Ephemerides and Nautical Almanacs 38

Physics 39

Chemistry 42

Geology 49

Geography 56

Other Earth Sciences 59
(a) Geodesy
(b) Meteorology
(c) Terrestrial Magnetism
(d) Oceanography

Natural History 61

Microscopy 68

Botany 68

Zoology 74
(a) General Zoology
(b) Entomology
(c) Ornithology

(d) Malacology and Conchology
(e) Ichthyology
(f) Marine Zoology

Experimental Biology 84

Microbiology 89

PART 2. SELECTION OF THE PERIODICALS AND THEIR RELATIVE SIZES

2.1 The Seventeenth and Eighteenth Centuries 91

2.2 The Nineteenth Century 101

PART 3. THE DEVELOPMENT OF THE PERIODICAL LITERATURE

3.1 Science in General 115

3.2 The Individual Sciences 135

Appendix: The Periodical Literature of the Individual Sciences at the Beginning of the Twentieth Century 171

Bibliography 173

Index 179

INTRODUCTION

This work is a companion to my *Historical Catalogue of Scientists and Scientific Books* and, like it, is intended to serve both as a general reference work and as a research tool for historians of science. It contains a list of some nine hundred scientific periodicals selected from a much greater number from the period 1665–1900 and arranged chronologically within subject groups. The second and third parts of the book include various analyses and an account of the evolution of the periodical literature of science as evidenced by the kinds and numbers of periodicals in the chronological list.

In preparing the list the fundamental issue was the problem of selection. Bibliographical information about periodicals of all historical periods is readily available from standard sources and, though laborious, it would not be difficult to compile a comprehensive list of periodicals from the period 1665–1900, especially as there are two valuable nineteenth-century bibliographies of scientific periodicals, namely those of Scudder (1879) and Bolton (1897). Such a list, even when restricted to the pure sciences, would contain several thousand items, a large proportion of which would be periodicals which ran for only a few years. If one decided to exclude all those which lasted less than some arbitrary period, say ten years, the list would be greatly reduced, but it would still contain a large proportion of periodicals of very little significance.

In general, the significance of a periodical for the historian is best assessed, at least as a first step, by the degree to which it was used as a means of publication by the scientists of the time, and this is the criterion that was adopted for selection of items to be included in the Catalogue. (In a few special cases other criteria were used, especially established historical importance.) In the case of the nineteenth-century periodicals, which constitute the vast majority, the degree of use was ascertained by means of surveys utilizing the two major periodical indexes, the Royal Society's *Catalogue of Scientific Papers* and the *International Catalogue of Scientific Literature*. These surveys yielded a measure of the relative sizes of the periodicals in terms of the numbers of papers cited, and the items included in the Catalogue were those above a certain minimum size. For the seventeenth- and eighteenth-century periodicals an approach of this kind was not suitable and a different method was adopted using the bibliographical data contained in Poggendorff's well-known *Handwörterbuch*. The selection of the periodicals and their relative sizes are discussed in detail in Part 2.

Part 3, which can be read independently of the somewhat technical discussion in Part 2, is a sketch of the development of scientific periodicals based on the data contained in Part 1 and on the published literature dealing with the subject. It is not intended to be a full history of scientific periodicals, and it says little about some important aspects, notably refereeing and the sociological functioning of periodicals in the scientific community. The time is not yet ripe for a comprehensive history of the subject. D.A. Kronick's *History of Scientific and Technical Periodicals, 1665–1790* (1976) is useful for the period it covers, but it is written from a librarian's viewpoint, being chiefly concerned with the forms of scientific journalism (many of the periodicals it mentions were of little or no significance for the scientific movement). Many more studies of individual periodicals need to be made; indeed some of the most important have not yet received any attention from historians. In addition there is much information to be gleaned from the history of scientific societies and other institutions as well as from the biographies of editors (many of whom were important scientists).

Some remarks should be made here about a theme that runs through the book, namely the two categories of periodicals and the terminology that is used for them. The first category, the institutional periodicals, comprises all those which are issued by institutions of any kind—scientific societies and academies, universities, museums, observatories, etc. These are invariably subsidized financially by their parent institution. The second category, sometimes called independent or proprietary but here called non-institutional, are those published on a commercial basis by a publisher in conjunction with an editor. In contrast to the institutional periodicals, these have to pay their way and if they fail to do so they go out of publication. In the past (to say nothing of the present) survival was difficult and there is an almost endless number of non-institutional periodicals which began hopefully but were forced to cease publication after a short period. Some, however, not only survived but flourished, and they include several of the most important periodicals in the history of science.

Because of the greater prominence of academies and societies in the scientific world of the past, there was formerly a greater awareness of the distinction between the two categories than there is today, and the term "journal" was generally used to designate the non-institutional periodicals. It is still sometimes used in this sense today but more often it is simply a synonym of "periodical." Consequently the term has not been used in contexts where it could cause confusion.

I am indebted to Dr. J. Gascoigne and Dr. D.R. Oldroyd for their helpful comments on the manuscript.

R.M.G.
School of History and Philosophy of Science
University of New South Wales
Sydney

LIST OF TABLES

1. Proportion of Scientists Publishing in Periodicals 92
2. The Chief Periodicals Beginning Before 1790 93
3. Examples of Short-lived Scientific Journals Excluded from the Catalogue 95
4. Examples of General Learned Journals Excluded from the Catalogue 96
5. Specialized Periodicals Beginning Before 1790 99
6. Contents of the Natural Science Section of Reuss' *Repertorium* 100
7. Categories of Size 102
8. Ranking by Size, 1863–64. Science in General 103
9. Ranking by Size, 1863–64. Mathematics 104
10. Ranking by Size, 1863–64. Astronomy 104
11. Ranking by Size, 1863–64. Physics 104
12. Ranking by Size, 1863–64. Chemistry 105
13. Ranking by Size, 1863–64. Geology 105
14. Ranking by Size, 1863–64. Geography 105
15. Ranking by Size, 1863–64. Meteorology 106
16. Ranking by Size, 1863–64. Natural History 106
17. Ranking by Size, 1863–64. Botany 106
18. Ranking by Size, 1863–64. Zoology 106
19. Ranking by Size, 1863–64. Experimental Biology 107
20. Ranking by Size, 1901–05. Science in General 107
21. Ranking by Size, 1901–05. Mathematics 108
22. Ranking by Size, 1901–05. Astronomy 108
23. Ranking by Size, 1901–05. Physics 109
24. Ranking by Size, 1901–05. Chemistry 109
25. Ranking by Size, 1901–05. Geology 110
26. Ranking by Size, 1901–05. Geography 110
27. Ranking by Size, 1901–05. Meteorology 111
28. Ranking by Size, 1901–05. Natural History 111
29. Ranking by Size, 1901–05. Microscopy 112
30. Ranking by Size, 1901–05. Botany 112
31. Ranking by Size, 1901–05. Zoology 112

32. Ranking by Size, 1901–05. Experimental Biology 113
33. Ranking by Size, 1901–05. Microbiology 114
34. General Science Periodicals Published by Institutions, 1665–1900 128
35. General Science Periodicals in Existence at Successive Times 132
36. Proportion of General Science Periodicals at Successive Times 133
37. Proportion of Institutional Periodicals 136
38. Proportion of Small Perodicals in 1901–05 136
39. Percentages of the Periodicals for Each Science Published in the Chief Countries 137
40. Periodicals in Existence at Successive Times. Mathematics 140
41. Periodicals in Existence at Successive Times. Astronomy 143
42. Periodicals in Existence at Successive Times. Physics 145
43. Periodicals in Existence at Successive Times. Chemistry 149
44. Periodicals in Existence at Successive Times. Geology 154
45. Periodicals in Existence at Successive Times. Geography 156
46. Periodicals in Existence at Successive Times. Natural History 157
47. Periodicals in Existence at Successive Times. Botany 160
48. Periodicals in Existence at Successive Times. Zoology 162
49. Periodicals in Existence at Successive Times. Experimental Biology 167
50. Average Number of References per Year in 1901–05 172

A Historical Catalogue of Scientific Periodicals 1665–1900

PART 1

THE CATALOGUE

Within each subject section the arrangement of the periodicals is chronological, according to their date of commencement. Only periodicals which began before 1900 are included; many of them, of course, continued into the twentieth century but name changes, amalgamations, etc., after 1900 are generally not recorded. The list is restricted to the pure sciences but, as is described more fully at the beginning of several of the sections, some periodicals of an applied or practical nature were included when they contained significant numbers of papers bearing on one or other of the pure sciences.

The titles were collected in the first place from the catalogues of Scudder (1879) and Bolton (1897) and from various other bibliographical sources, and secondly from the three surveys described in Part 2. Except for special cases, periodicals which ran for less than ten years were not included; a very large number was thereby excluded. The methods of selecting those to be included are discussed in Sections 2.1 and 2.2. The categories used to represent the sizes of the nineteenth-century periodicals are explained in Section 2.2 (Table 7 and accompanying discussion).

The date, followed by a plus sign, which appears in each entry after the title is the date of the first volume. If the date of publication of the first volume is known to be different from the year or years which the volume covers, then the date of publication is given in parentheses. Thus "1872/73 (1874)+" means that the first volume covered the years 1872 and 1873 and was published in 1874. However the various bibliographical sources used were often unclear and sometimes conflicting about such dates, consequently they may be in error by a year or two. No information is given (except in a few special cases) about volume numbering, series, or indexes. If required, such information can be obtained from the catalogues of Scudder and Bolton or from standard bibliographical sources (especially the *British Union-Catalogue of Periodicals* which lists periodicals by their original names; the American *Union List of Serials*, on the other hand, lists them by their latest names but it sometimes includes useful historical notes; both these catalogues give library locations, as do the *World List of Scientific Periodicals* and various national union-catalogues). Place of publication is not given if it is the same as the place of the issuing body. In the case of non-institutional periodicals the name of the first editor, who was generally the founder, is given when known.

The entry numbers all begin with 0 in order to distinguish them from the page numbers, especially in the Index. At the end of each entry references are given, where applicable, to the catalogues of Scudder and Bolton. The annotation "See Index" means that there is a discussion of the particular periodical in Parts 2 or 3 which can be found by means of the Index.

The term 'periodical' has been interpreted liberally and a few items are included which are not periodicals in the usual sense (e.g. 022, 0682, 0691).

SCIENCE IN GENERAL

01. *Journal des sçavans* (from 1793 *savans*). Paris, 1665+. Edited initially by D. de Sallo, then by J. Gallois from 1666 to 1675, and subsequently by various individuals until 1702 when a panel of editors was appointed by the French government.
See Index. Scudder 1424, 1425. Bolton 2449. Documentation: Birn, 1965; Brown, 1934, 1972; Morgan, 1929; Paris, 1903; Sergescu, 1936, 1947; DSB--articles on Sallo and Gallois.

02. *Philosophical transactions*. London, 1665+. Edited by H. Oldenburg until his death in 1677 and thereafter by successive secretaries of the Royal Society until 1753 when its publication was taken over by a committee of the Society. Size: 1800-63, 3; 1864-73, 3; 1901-05, 6.
See Index. Scudder 434g. Documentation: Andrade, 1965; George, 1952; Hall, 1965, 1975; Wightman, 1961.

03. Academia Naturae Curiosorum. (Founded 1652) From 1682 Academia Caesareo-Leopoldina Naturae Curiosorum. Later Kaiserlich Leopoldinisch-Carolinische Academie der Naturforscher. Now Deutsche Akademie der Naturforscher. See also 0705.
Miscellanea curiosa medico-physica. Leipzig, etc., 1670-1706. Cont'd as *Ephemerides*. Frankfurt, etc., 1712-22. Cont'd as *Acta physico-medica*. Nürnberg, 1727-54. Cont'd as *Nova acta physico-medica*. Nürnberg, etc., 1757+. (Also entitled *Verhandlungen* from 1818) Size: 1818-64, 1; 1865-73, 1; 1901-05, 1.
See Index. Scudder 2808, 2580, 3146. Bolton 3051, 29. Documentation: Röpke, 1928.

04. *Acta eruditorum*. Leipzig, 1682+. Edited by O. Mencke until his death in 1707. Cont'd from 1732 as *Nova acta eruditorum*. Ceased 1776 (1782). (The last editor, K.A. Bel, died in 1782.)
See Index. Scudder 2880. Bolton 24, 25. Documentation: Loria, 1941.

05. Académie Royale des Sciences. Paris. (Founded 1666) See also 010 and 018.
Histoire.... Avec les mémoires de mathématique et de physique. 1699 (1702)+. Ceased 1789 (1793). (The volume for 1790 was published in 1797.) For its successor see 047.
See Index. Scudder 1276d.
The memoirs submitted to the Academy by its members in the period 1666-1699 were, with some exceptions, not published by it at the time (though brief accounts of many of them appeared in the

Journal des sçavans or similar journals). They were published in full, mostly for the first time, in an eleven-volume collection (not a periodical) issued by the Academy in 1729-33. See Kronick (1976, pp. 142-143) and Russo (1969, pp. 54-55).

06\. *Mémoires pour servir à l'histoire des sciences et des beaux-arts.* Trévoux, 1701+. Cont'd from 1768 as *Journal des sciences et des beaux arts.* Cont'd from 1779 as *Journal de littérature, des sciences et des arts.* Ceased 1782. (Generally referred to as *Mémoires de Trévoux* or *Journal de Trévoux*)
See Index. Scudder 1690, 1428, 1212. Bolton 2924. Documentation: Desautels, 1956; Dumas, 1936.

07\. Societas Regia Scientiarum. Berlin. (Founded 1700) See also 017.
Miscellanea Berolinensia ad incrementum scientiarum. 1710-43 (1746) For its successor see 017.
See Index. Scudder 2385. Bolton 3049.
Vol. 1 was published in 1710 but because of the Academy's difficulties in its early years Vol. 2 did not appear until 1723 and Vol. 3 until 1727. Five more volumes were published in the period 1734-46.

08\. *Giornale de' letterati d'Italia.* Venezia, 1710+. Edited by A. Zeno. (Not to be confused with other journals of the same name) Ceased 1740.
See Index. Scudder 2095. Bolton 1953. Documentation: Loria, 1899.

09\. Académie Royale des Sciences, Belles-Lettres et Arts de Bordeaux. (Founded 1712)
Recueil des dissertations qui ont remporté le prix. 1715-35 (1741) 6 vols. (Cont'd 1740-93. 2 vols)?
See Index. Scudder 1061.

010\. Académie Royale des Sciences. Paris. See also 05.
Recueil des pièces qui ont remporté le prix. 1720-72 (1721-77). 9 vols.
See Index. Scudder 1276j. Documentation: Jaeggli, 1977.

011\. Regia Societas Scientiarum Upsaliensis. (Founded 1710)
Acta literaria (from 1730 *et scientiarum*) *Sueciae.* 1720+. Cont'd from 1740 as *Acta Societatis....* Cont'd from 1773 as *Nova acta Societatis....* Size: 1799-1863, 1; 1864-73. 1. Cont'd past 1900.
See Index. Scudder 704, 705, 711. Bolton 27.

012\. Academia Scientiarum Imperialis Petropolitana. St. Petersburg. (Founded 1724) See also 055.
Commentarii. 1726 (1728)+. Cont'd from 1747 (1750) as *Novi commentarii.* Cont'd from 1777 (1778) as *Acta.* Cont'd from 1783 (1787) as *Nova acta.* Ceased 1802 (1806); for its successor see 055.
See Index. Scudder 3706.

013\. Scientiarum et Artium Institutum Bononiense atque Academia. (Founded 1714) See also 090.
Commentarii. 1731+. Cont'd from 1834 as *Novi commentarii.* Size: 1834-49, 2. Ceased 1849; for its successor see 0140.
See Index. Scudder 1807, 1796.

014\. Svenska Vetenskaps Academien. Stockholm. (Founded 1739) See also 074.

(a) *Handlingar*. 1739+. Cont'd from 1780 as *Nya handlingar*. Size: 1800-12, 2. Cont'd from 1813 as *Handlingar*. Size: 1813-56, 1; 1857-73, 1; 1901-05, 2.
See Index. Scudder 693u.

(b) ——— [German translation] Kön. Schwedische Akademie der Wissenschaften.
Abhandlungen. 1739 (1749)+. Trans. with commentary by Holzbecher and A.E. Kästner. Ceased 1791 (1794).
Scudder 2768. Bolton 14.

015. *Göttingische Zeitungen von gelehrten Sachen*. 1739+. Edited by A. von Haller from 1747 to 1753. Cont'd from 1753 as *Göttingische Anzeigen von gelehrten Sachen*. Published by the Kön. Gesellschaft der Wissenschaften in Göttingen. Cont'd from 1802 as *Göttingische gelehrte Anzeigen*. Cont'd past 1900.
See Index. Scudder 2685, 2678c,d. Bolton 2005. Documentation: Kronick, 1976, p. 185.

016. Kjøbenhavnske Selskab af Laerdoms og Videnskabers Elskere. [Copenhagen Society of Friends of Learning and Science] (Founded 1742) From 1777 K. Danske Videnskabers Selskab; see 052.
Skrifter. 1743 (1745)+. Cont'd from 1781 as *Skrifter: Nye samling*. Ceased 1799; for its successor see 052.
See Index. Scudder 622, 699, 615.

017. Académie Royale des Sciences (*later* et des Belles-Lettres). Berlin. (For its earlier period see 07; for its later period see 056)
Histoire ... avec les mémoires. 1745 (1746)+. Cont'd from 1770 (1772) as *Nouveaux mémoires*. Cont'd from 1786 (1792) as *Mémoires*. Size: 1799-1804, 4. Ceased 1804 (1807); for its successor see 056.
See Index. Scudder 2262.

018. Académie Royale des Sciences. Paris. See also 05.
Mémoires de mathématique et de physique, présentés à l'Académie ... par divers sçavans et lûs dans ses assemblées. 1750-86. (Often referred to as *Mémoires par savans étrangers*)
See Index. Scudder 1276g.

019. *Acta Helvetica, physico-mathematico-(anatomico-) botanico-medica*. Basel, 1751+. Published by the Societas Physico-Medica Basiliensis. (A typical learned society's periodical despite the rather unusual form of the title.) Ceased 1777. A single volume with the title *Nova acta Helvetica* appeared in 1787.
See Index. Scudder 2128. Bolton 26.

020. Societas Regia Scientiarum Gottingensis. (Founded 1751) See also 0105.
Commentarii. 1751 (1752)+. Cont'd from 1758 (1763) as *Commentationes*. Cont'd from 1769 (1771) as *Novi commentarii*. Cont'd from 1778 (1779) as *Commentationes*. Cont'd from 1808 (1811) as *Commentationes recentiores*. Ceased 1837 (1838); for its successor see 0105.
See Index. Scudder 2694, 2434.

021. *Commentarii de rebus in scientia naturali et medicina gestis*. Leipzig, 1752+. Edited by C.G. Ludwig, J.S. Reichel, et al.

Ceased 1808.
See Index. Scudder 2913. Bolton 1325.
"The first scientific review journal of any extent and duration .... It appeared in quarterly issues of about two hundred pages each, and contained long reviews of scientific books, dissertations and journals, especially society proceedings. It had the advantage that it was printed in Latin and was therefore accessible to the entire learned world. The journal articles are abstracted at some length ... it also contained short sections which provided reports of academic news and lists of [new] books.... Each volume contained its own author and subject index and decennial indexes were issued to the first thirty volumes.... It was widely known and honored." --Kronick, 1976, p. 188.

022. [*Collection académique*, in two parts]
(a) [Partie française] *Recueil de mémoires; ou, Collection des pièces académiques, concernant la médecine, l'anatomie et la chirurgie, la chymie, la physique expérimentelle, la botanique et l'histoire naturelle.* Dijon/Paris, 1754-87. 16 vols.
(b) [Partie étrangère] *Collection académique, composée des mémoires, actes, ou journaux des plus célèbres académies et sociétés littéraires étrangères, des extraits des meilleurs ouvrages périodiques ... concernant l'histoire naturelle et la botanique, la physique expérimentelle et la chymie, la médecine et l'anatomie, traduits en françois.* Dijon/Paris, 1755-79. 13 vols.
See Index. Scudder 1139, 1141. Bolton 1310, 3884. Documentation: Kronick, 1976, pp. 212-216.
Not really a periodical but a type of retrospective collection characteristic of the eighteenth century. The "Partie française" consists of abridgements of the *Mémoires* of the Académie des Sciences, while the "Partie étrangère" comprises numerous long extracts, translated into French, from the leading periodicals of other countries. The period covered extends from 1665 and the coverage is actually better for the late seventeenth century than for the eighteenth. Mathematics is explicitly excluded. The founder of this ambitious project was Jean Berryat (died 1754); later editors included Buffon, Daubenton, and other leading French scientists of the time.

"The *Collection académique* does not provide a very complete repository of the scientific advances of the 17th and 18th centuries. It does, however, provide an extremely valuable key to a large part of the literature of this period. It is made even more valuable by the four-volume index to both series which was issued between 1775 and 1776 by the Abbé Rozier. Rozier's index also provides a key to the various publications of the Académie des Sciences de Paris to the end of 1770." --Kronick, 1976, p. 215. (Kronick goes on to describe Rozier's index in some detail.)

023. Hollandsche (*sometimes* Bataafsche) Maatschappij der Wetenschappen te Haarlem. (Société Hollandaise des Sciences) (Founded 1752) See also 0197.
Verhandelingen. 1754+. Cont'd from 1799 as *Natuurkundige ver-*

handelingen. Size: 1799-1844, 1; 1841-68, 1. Ceased 1922.
See Index. Scudder 849, 847.

024. Academia Electoralis Moguntina Scientiarum Utilium quae Erfurti est. (Founded 1754) Later Kurfürstlich-Mainzische Akademie Nützlicher Wissenschaften zu Erfurt.
Acta. 1757+. Cont'd from 1799 as *Nova acta* (also entitled *Abhandlungen*). Ceased 1826. Some fragmentary publications later in the 19th century.
See Index. Scudder 2546, 2547.

025. *Miscellanea philosophico-mathematica societatis privatae Taurinensis*. 1759. Cont'd from 1760 as *Mélanges de philosophie et de mathématique de la Société royale de Turin*. Ceased 1773; for its successor see 038. The society was a precursor of the Turin Academy--038.
See Index. Scudder 2068, 2075. Bolton 3058.

026. Accademia delle Scienze di Siena detta de' Fisiocritici. (Founded 1691)
Atti. 1761+. Size: 1808-41, 1; 1901-05, 1.
See Index. Scudder 2050.

027. Churfürstlich-Baierische (later Königliche Baierische) Akademie der Wissenschaften. München. (Founded 1759)
Abhandlungen. 1763+. Cont'd from 1778 as *Neue philosophische Abhandlungen*. Ceased 1797. For later publications see 063.
See Index. Scudder 3114, 3105, 3109.

028. Academia Electoralis Scientiarum et Elegantiorum Literarum Theodoro-Palatina. Mannheim. (Founded 1775)
Historia et commentationes (also entitled *Acta physica et historica*). 1766-94. No further periodicals.
See Index. Scudder 3067.

029. American Philosophical Society. Philadelphia. (Founded 1744; re-founded 1769) See also 0106.
Transactions. 1769 (1771)+. (Suspended 1809-1818) Size: 1771-1809, 2; 1818-63, 1; 1864-73, 1; 1901-05, 1.
See Index. Scudder 4240.

030. *Introduction aux observations sur la physique, sur l'histoire naturelle et sur les arts*. Paris, 1771+. Cont'd from 1773 as *Observations sur....* Cont'd from 1794 as *Journal de physique, de chimie, et d'histoire naturelle*. (For the period before 1794 it is sometimes referred to retrospectively as *Journal de physique*.) Edited by F. Rozier until 1780, then by J.A. Mongez until 1785, then by J.C. de Lamétherie until his death in 1817, and finally by H.M. Ducrotay de Blainville. Size: 1800-23, 4. Ceased 1823.
See Index (under Rozier). Scudder 1398, 1607, 1421. Bolton 2201. Documentation: Kronick, 1976, pp. 106-112; McClellan, 1979; McKie, 1957; Neave, 1950-52; DSB--article on Lamétherie.

031. Bataafsch Genootschap der Proefonderwindelijke Wijsbegeerte. [Batavian Society of Experimental Philosophy] Rotterdam. (Founded 1769)
Verhandelingen. 1774+. Cont'd from 1800 as *Nieuwe verhandelingen*. Size: 1800-51, 1. Cont'd past 1900.
See Index. Scudder 898.

032. *Scelta di opuscoli interessanti, tradotti da varie lingue.* Milano, 1775+. Edited by C. Arnozett et al. Cont'd from 1778 as *Opuscoli scelti, sulle scienze e sulle arti.* Cont'd from 1804 as *Nuova scelta d'opuscoli interessanti.* Ceased 1807.
See Index. Scudder 1912, 1903. Bolton 4211.

033. *Abhandlungen einer Privatgesellschaft in Böhmen.* Prag, 1775-84. Edited by I.E. von Born. For continuation see 040.
See Index. Scudder 3482.

034. Académie Impériale et Royale des Sciences et Belles-Lettres de Bruxelles. (Founded 1772)
Mémoires. 1777-88. (The volume for 1789 was published in 1820) For its successor see 073.
See Index. Scudder 943.

035. *Magazin für das Neueste aus der Physik und Naturgeschichte.* Gotha, 1781+. Edited by L.C. Lichtenberg until 1786, then by J.H. Voigt. Cont'd from 1797 as *Magazin für den neuesten Zustand der Naturkunde.* Edited by Voigt. Size: 1797-1806, 2. Ceased 1806.
See Index. Scudder 2666. Bolton 2775, 2776.

036. Académie Royale des Sciences, Inscriptions et Belles-Lettres. Toulouse. (Founded 1640; re-founded 1729) See also 087.
Histoire et mémoires. 1782-90. For continuation see 087.
See Index. Scudder 1679.

037. Società Italiana dei Quaranta (*or* dei XL). (Founded in Verona in 1782 by the mathematician and physicist, A.M. Lorgna) From 1804 Società Italiana delle Scienze (residente in Modena).
Memorie di matematica e di fisica. Verona, later Modena, 1782+. Size: 1782-1855, 1; 1862-73, 1. Cont'd past 1900.
See Index. Scudder 1924, 2108.

038. Académie Royale des Sciences de Turin. (Founded 1783) See also 065 and 0189.
Mémoires (continuation of 025). 1784 (1786)+. Size: 1784-1815, 1. For continuation after 1815 see 065.
See Index. Scudder 2051, 2052.

039. Manchester Literary and Philosophical Society. (Founded 1781) See also 0167.
Memoirs. 1785+. Size: 1789-1802, 1; 1805-60, 1; 1862-73, 1; 1901-05, 2.
See Index. Scudder 494.

040. K. Böhmische Gesellschaft für Wissenschaften. Prag. (Founded 1784. Precursor: see 033) See also 0170.
Abhandlungen. 1785 (1786)+. Size: 1804-22, 1; 1837-64, 1. Ceased 1892.
See Index. Scudder 3492.

041. American Academy of Arts and Sciences. Boston. (Founded 1780) See also 0127.
Memoirs. 1780/85 (1785)+. (Suspended 1818(1821)-1833) Size: 1785-1818, 1; 1833-63, 1. Cont'd past 1900.
See Index. Scudder 3974.

042. Cesareo-Regia (*later* Reale) Accademia di Scienze, Lettere ed Arti

di Padova. (Founded 1540; re-founded 1779)
Saggi. 1786-94. Resumed in 1817 as *Nuovi saggi*. Size: 1817-63, 1. Cont'd from 1884 as *Atti e memorie*. Size: 1901-05, 1.
See Index. Scudder 1964.

043. Royal Irish Academy. Dublin. (Founded 1785) See also 0102.
Transactions. 1787+. Size: 1787-1859, 1; 1901-05, 1. Ceased 1906.
See Index. Scudder 93.

044. Royal Society of Edinburgh. (Founded 1783) See also 0120.
Transactions. 1783/88 (1788)+. Size: 1788-1864, 1; 1865-73, 2; 1901-05, 2.
See Index. Scudder 124.
The following was a predecessor: *Essays and observations, physical and literary, read before a society in Edinburgh and published by them*. 1754-65 (2nd ed., 1770-71), 3 vols. The society included W. Cullen and J. Black.
Scudder 117. Bolton 1636. (German translation: Scudder 2229; Bolton 4674. French translation: Scudder 1384; Bolton 1635)

045. Société Philomatique de Paris. (Founded 1788)
Bulletin des sciences. 1791+. Size: 1791-1805, 2. Cont'd from 1807 as *Nouveau bulletin....* Size: 1807-25, 4. Suspended 1827-31 and 1834-35. Cont'd from 1836 (1837) as *Extrait des procès-verbaux des séances*. Size: 1836-52, 5. Cont'd from 1864 as *Bulletin*. Size: 1901-05, 1.
See Index. Scudder 1594 & errata.

046. Ecole Centrale des Travaux Publics. Paris. (Founded 1794) From 1795 Ecole Polytechnique.
Journal polytechnique. 1794. Cont'd from 1795 as *Journal de l'Ecole polytechnique*. Size: 1795-1863, 1; 1864-73, 1; 1901-05, 1. Chiefly mathematical.
See Index. Scudder 1377, 1379. Bolton 2334.

047. Institut National de France. Paris. (Founded 1795. Successor to the Académie Royale des Sciences and the other pre-revolutionary academies.) See also 057.
Mémoires: Sciences mathématiques et physiques. 1796 (1798) - 1815 (1818). Size: 1796-1815, 2. For its successor see 066.
See Index. Scudder 1396.

048. (a) *Bibliothèque britannique, ou recueil extrait des ouvrages anglais, périodiques et autres.... Partie des sciences et arts*. Genève, 1796+. Edited by F.G. Maurice, C. Pictet, and M.A. Pictet. Size: 1800-15, 1. Cont'd as (b).
(b) *Bibliothèque universelle des sciences, belles-lettres et arts .... [Partie des] Sciences et arts*. Ib., 1816+. Size: 1816-35, 4. Cont'd as (c).
(c) *Bibliothèque universelle de Genève*. Ib., 1836+. (The separate parts were abolished.) Edited by A.A. de La Rive. Size: 1836-45, 8. For continuation after 1845 see 0129.
See Index. Scudder 2154. Bolton 933. Documentation; Bickerton, 1972.

049. *A journal of natural philosophy, chemistry, and the arts*. London, 1797+. Edited by W. Nicholson. Size: 1797-1801, 11; 1802-13, 11. In 1814 it was incorporated in the *Philosophical magazine* (050).

See Index (under Nicholson). Scudder 324. Bolton 2518. Documentation: Lilley, 1948; DSB--article on Nicholson.

050. *The philosophical magazine.* 1798+. Edited by A. Tilloch (see below). In 1814 it incorporated Nicholson's Journal (049) and cont'd as *The philosophical magazine and journal.* In 1827 it incorporated *Annals of philosophy* (064) and cont'd as *The philosophical magazine; or, Annals of chemistry, mathematics, astronomy, natural history, and general science.* In 1832 it incorporated *The Edinburgh journal of science* (083) and cont'd as *The London and Edinburgh philosophical magazine and journal of science.* Cont'd from 1840 as *The London, Edinburgh, and Dublin philosophical magazine and journal of science* (known to the present day as simply *The philosophical magazine*). Founded by A. Tilloch and edited by him until his death in 1825, then by R. Taylor until 1851, and thereafter for some decades by W. Francis. From the 1830s its editorial board included some outstanding scientists, mostly physicists and chemists. Size: 1798-1826, 10; 1827-50, 8; 1851-63, 9; 1864-73, 12; 1901-05, 11 (physics 57%, chemistry 22%, mechanics 8%, mathematics 4%, other 8%).
See Index. Scudder 412. Bolton 3603. Documentation: Barr, 1964; Ferguson, 1948; DSB--article on Tilloch.

051. The Dublin Society. (Founded 1731; concerned chiefly with agriculture and technology) See also 0163.
Transactions. 1799(1800)-1810. Size: 1799-1810, 1.
Scudder 85.

052. K. Danske Videnskabers Selskab. Kjøbenhavn. See also 016, 059, and 082.
Skrivter. (Continuation of 016) 1800 (1801)+. Size: 1801-18, 1; 1849-73, 1; 1901-05, 1.
Scudder 615.

053. Société d'Amateurs des Sciences et Arts (*from 1827* Société des Sciences et Arts) de Lille.
Mémoires. 1802+. Size: 1827-53, 1. Cont'd past 1900.
Scudder 1165, 1167.

054. Société des Sciences et Belles-lettres de Nancy. (Name varies. Its publication record dates from 1754.) *From 1852* Académie de Stanislas.
Précis analytique des travaux. 1802 (1803)+. Size: 1805-32, 1. Cont'd from 1833 (1835) as *Mémoires.* Size: 1833-51, 2; 1852-63, 1. Cont'd past 1900.
Scudder 1248, 1243.

055. Académie Impériale des Sciences de St. Pétersbourg. See also 012, 0101, and 0801.
Mémoires. 1803 (1809)+. Size: 1803-30, 2; 1831-59, 2; 1859-73, 2; 1901-05, 2.
Scudder 3707.

056. K. (*later* K. Preussische) Akademie der Wissenschaften zu Berlin. See also 017 and 0103.
Abhandlungen. 1804/11 (1815)+. Size: 1804-63, 1; 1864-73, 1; 1901-05, 1.
Scudder 2370a.

057. Institut National de France. See also 047.
Mémoires présentés à l'Institut par divers savans et lûs dans ses assemblées: Sciences mathématiques et physiques. 1805-11. (Successor to 018. Generally known as *Mémoires des savans étrangers*) Size: 1805-11,1. For its successor see 088.
Scudder 1396.

058. Société d'Agriculture, d'Histoire Naturelle et Arts Utiles de Lyon.
Compte rendu. 1806+. Cont'd from 1825 as *Mémoires.* Cont'd from 1838 as *Annales des sciences physiques et naturelles, d'agriculture et d'industrie.* Size: 1838-48, 1; 1849-61, 2; 1864-73, 1. Cont'd past 1900.
Scudder 1188.

059. K. Danske Videnskabers Selskab. Kjøbenhavn. See also 052 and 082.
Oversigt over det ... forhandlinger. 1806(?)+ (or 1815(?)+) Size: 1832-61, 2; 1864-73, 1; 1901-05, 2.
Scudder 615.

060. Société d'Arcueil. (Founded 1807 at Arcueil, near Paris)
Mémoires de physique et de chimie. Paris, 1807-17. 3 vols. (It included some important papers in other fields besides physics and chemistry.) Size: 1807-17, 2.
See Index. Scudder 1558. Documentation: Crosland, 1967.

061. Schlesische Gesellschaft für Vaterländische Cultur. Breslau. (Founded 1803)
Correspondenzblatt. 1807+. Cont'd from 1824 as *Übersicht der Arbeiten....* Cont'd from 1850 as *Jahresbericht.* Size: 1835-63, 6; 1864-73, 2; 1901-05, 3 (mostly physical and biological sciences).
Scudder 2445, 2456.

062. *Giornale di fisica, chimica e storia naturale.* Pavia, 1808+. Edited by L. Brugnatelli. Size: 1808-17, 4; 1818-27, 3. Ceased 1827.
Scudder 1992. Bolton 1972.

063. K. Bayerische Akademie der Wissenschaften. München. See also 027 and 097.
Denkschriften. 1808 (1809)+. Size: 1809-24, 1. Cont'd from 1829/30 (1832) as *Abhandlungen der math.-phys. Classe.* Size: 1832-63, 1; 1864-73, 1; 1901-05, 1.
Scudder 3109.

064. *Annals of philosophy; or, Magazine of chemistry, mineralogy, mechanics, natural history, agriculture, and the arts.* London, 1813+. Edited by T. Thomson; from 1821 by R. Phillips. Size: 1813-20, 8; 1821-26, 5. In 1827 it was incorporated in *The philosophical magazine* (050).
See Index. Scudder 194. Bolton 427.

065. R. Accademia delle Scienze di Torino. See also 038 and 0189.
Memorie. (Continuation of 038) 1815 (1818)+. Size: 1816-38, 1; 1839-65, 2; 1901-05, 1.
Scudder 2053.

066. Académie (Royale) des Sciences. Paris. (Re-established in 1816

as a largely independent section of the Institut de France) See also 088 and 098.
Mémoires. (Successor to 047) 1816 (1818)+. Size: 1816-61, 1; 1864-73, 2; 1901-05, no references found.
See Index. Scudder 1397c.

067. Royal Institution of Great Britain. (Founded 1799) See also 0145.
The journal of science and the arts. 1816+. Cont'd from 1819 as *The quarterly journal of literature, science and the arts*. Size: 1816-27, 6; 1827-30, 9. Cont'd from 1830 as *Journal of the Royal Institution....* Ceased 1831.
See Index. Scudder 309 & errata. Bolton 2524 & p. 803.

068. *Biblioteca italiana; ossia, Giornale di letteratura, scienze ed arti*. Milano, 1816+. Size: 1816-40, 2. Succeeded in 1841 by 0111.
Scudder 1884. Bolton 923.

069. *The American journal of science (and arts)*. New Haven, Conn., 1818+. Edited by B. Silliman, Snr., then from the 1840s by several joint editors including B. Silliman, Jnr., and J.D. Dana, and later E.S. Dana. Size: 1818-45, 7; 1846-63, 11; 1864-73, 11; 1901-05, 11 (chemistry 28%, mineralogy 15%, physics 14%, zoology 12%, palaeontology 12%, geology 8%, other 11%).
See Index (also under Silliman). Scudder 4153. Bolton 223 & p. 619. Documentation: Dana, 1918; DSB--article on Silliman.

070. R. Accademia delle Scienze e Belle Lettere di Napoli. (Founded 1756. In the 18th century it published only one volume of *Atti*. In 1817 it became a section of the Società Reale Borbonica.) See also 0115 and 0174.
Atti. 1819+. Size: 1819-51, 1. Ceased 1851; for its successor see 0174.
Scudder 1933, 1934.

071. *The Edinburgh philosophical journal*. Edinburgh, 1819+. Edited by D. Brewster and R. Jameson. Size: 1819-26, 4. Cont'd from 1826 as *The Edinburgh new philosophical journal*. Size: 1826-54, 6; 1855-64, 4. In 1864 it was incorporated in the *Quarterly journal of science* (0183).
See Index. Scudder 115, 116. Bolton 1528.

072. *Giornale arcadico di scienze, lettere ed arti*. Roma, 1819+. Size: 1819-56, 1; 1857-60, 1. Ceased 1909.
Scudder 2032. Bolton 1929.

073. Académie Royale des Sciences et Belles-Lettres de Bruxelles. See also 034 and 093.
Nouveaux mémoires. (Continuation of 034) 1820+. Cont'd from 1847 as *Mémoires*. Size: 1820-64, 1; 1901-04, 1. Ceased 1904.
Scudder 943.

074. K. Vetenskaps Akademien. Stockholm. See also 014, 0118, and 0216.
Årsberättelser om vetenskapernas framsteg. [*Annual accounts of the progress of science*] 1820 (1821)+. Cont'd from 1826 in a number of separate series for the various individual sciences. All these series ceased by the 1850s. A few of them were translated into German, the most notable being those for chemistry (0401) and botany (0686).
See Index. Scudder 693. Bolton 651-660, 860.

075. Naturforschende Gesellschaft. Danzig. (Founded 1743; fragmentary publication record from 1747)
Neueste Schriften. 1820+. Size: 1820-56, 1. Cont'd from 1863 as *Schriften.* Size: 1864-73, 1; 1901-05, 2. Ceased 1938.
Scudder 2485.

076. *Notizen aus dem Gebiete der Natur- und Heilkunde.* Erfurt, 1821+. Edited by L.F. von Froriep until his death in 1847, then by M.J. Schleiden and R. Froriep. Size: 1822-36, 8; 1837-46, 10; 1847-49, 11. Cont'd from 1850 as *Tagsberichte über die Fortschritte der Natur- und Heilkunde.* Ceased 1852.
Scudder 2559, 3310. Bolton 3368.
Though it covered the whole range of the sciences it had a strong biological emphasis as well as containing much medicine.

077. Cambridge Philosophical Society. (Founded 1819) See also 0193.
Transactions. 1821 (1822)+. Size: 1822-58, 1; 1901-05,1.
Scudder 61.
Apparently it was largely supplanted by the *Proceedings* (0193) after 1858.

078. Société de Physique et d'Histoire Naturelle de Genève.
Mémoires. 1821+. Size: 1821-63, 1; 1864-73, 1; 1901-05, 1.
Scudder 2163.

079. Gesellschaft Deutscher Naturforscher und Ärzte. (Founded 1822. A *Wandergesellschaft* which held annual meetings in different German cities.)
(Amtlicher) Bericht (later *Tageblatt* or *Verhandlungen*) *der ... Versammlung.* 1822+. (Not published separately until 1828; until then--and in a few later years--it was published in *Isis*--0754.) Size: 1822-52, 3; 1901-05, 11 (physiology 28%, chemistry 28%, anatomy 10%, botany 7%, zoology 6%, bacteriology 5%, other 16%)
See Index. Scudder 2220.

080. *Magazin for naturvidenskaberne.* Christiana, 1823+. Edited by G.F. Lundh et al. From 1832 published by the Physiographiske Forening i Christiana. Cont'd from 1838 as *Nyt magazin....* Size: 1838-63, 1; 1901-05, 2.
Scudder 583, 593. Bolton 2761.

081. *Bulletin universel des sciences et de l'industrie.* Edited by A.E.d'A. Férussac. Divided into eight sections, of which only the following two are relevant to the present catalogue.
See Index.
(a) *Bulletin des sciences mathématiques, astronomiques, physiques et chimiques.* Paris, 1824-31. Size: 1824-31, 2. Scudder 1596a. Bolton 1141. Documentation: Taton, 1947.
(b) *Bulletin des sciences naturelles et de géologie.* Paris, 1824-31. Size: 1824-31, 10. Scudder 1596b. Bolton 1144.

082. K. Danske Videnskabersselskab. Kjøbenhavn. See also 052.
Naturvidenskabelige og mathematiske afhandlinger. 1824+. Size: 1824-47, 1. Ceased 1938.
Scudder 615.

083. *The Edinburgh journal of science.* 1824+. Edited by D. Brewster.

Size: 1824-29, 6; 1829-32, 4. In 1832 it was incorporated in *The philosophical magazine* (050).
See Index. Scudder 114. Bolton 1527.

084. *Archiv für die gesammte Naturlehre*. Nürnberg, 1824-35. Edited by K.W.G. Kastner. Size: 1824-35, 5. Ceased 1835.
Scudder 3152, 3151. Bolton 603.

085. Accademia Gioenia di Scienze Naturali in Catania. See also 0255.
Atti. 1825+. Size: 1825-37, 1; 1844-62, 1; 1901-05, 2.
Scudder 1813.

086. (Allgemeine) Schweizerische Gesellschaft für die gesammten Naturwissenschaften. (Société Helvétique des Sciences Naturelles) (Società Elvetica delle Scienze Naturali) A *Wandergeschellschaft* which held annual meetings in different Swiss cities.
Verhandlungen ... Jahresversammlung. (In some years entitled *Actes* or *Atti*.) 1825+. (The annual meetings began in 1815 but the *Verhandlungen* were not begun until 1825.) Size: 1825-63, 2; 1864-73, 4; 1901-05, 4.
Scudder 2113n.

087. Académie des Sciences, Inscriptions et Belles-Lettres de Toulouse. (Founded 1729; publication record dates from 1782; re-founded 1807) See also 036.
Histoire et mémoires (continuation of 036) (later *Mémoires*). 1807/27 (1827)+. Size: 1827-44, 1; 1844-56, 3; 1857-63, 1; 1864-73, 1; 1901-05, 1.
Scudder 1679.

088. Académie (Royal) des Sciences. Paris. See also 066.
Mémoires presentés par divers savans (successor to 057). 1827+. Often referred to as *Mémoires des savans étrangers*. Size: 1827-62, 1; 1864-73, 1; 1901-05, no references found. Ceased 1914.
Scudder 1397.

089. Académie des Sciences, Belles-Lettres et Arts de Clermont-Ferrand.
Annales scientifiques, littéraires et industrielles de l'Auvergne. 1827+. Size: 1828-58, 1. Cont'd from 1859 as *Mémoires de l'Académie*. Cont'd past 1900.
Scudder 1130. Bolton 340.

090. Accademia delle Scienze dell'Istituto di Bologna. See also 013 and 0140.
Rendiconto. 1829+. Size: 1851-73, 1; 1901-05, 1.
Scudder 1798.

091. *Annali delle scienze del regno Lombardo-Veneto*. Padova, 1831+. Edited by A. Fusinieri. Size: 1831-45, 3. Ceased 1845.
Scudder 1966. Bolton 399.

092. British Association for the Advancement of Science. (Founded in 1831 to hold annual meetings in different British cities)
Report of the ... meeting. 1831+. Size: 1833-64, 11; 1864-73, 11; 1901-05, 11 (chemistry 24%, zoology 15%, botany 14%, geology 14%, mineralogy 9%, meteorology 8%, other 16%).
See Index. Scudder 22.

093. Académie Royale des Sciences et Belle-Lettres de Bruxelles. See

also 073.
Bulletins. 1832 (1834)+. Size: 1832-63, 4; 1864-73, 7; 1901-05, 1.
Scudder 943.

094. Royal Society of London. See also 02.
Abstracts of the papers printed in the "Philosophical transactions." 1800/14 (1832)+. Cont'd from 1843/50 (1851) as *Abstracts of the papers communicated to the Royal Society.* Cont'd from 1854/55 (1856) as *Proceedings of the Royal Society.* Size: 1832-63, 4; 1864-73, 10; 1901-05, 11 (physics 18%, chemistry 18%, astronomy 10%, zoology 9%, botany 8%, physiology 6%, meteorology 6%, other 25%)
See Index. Scudder 434a,j.
From the 1850s the *Proceedings* became the Society's main periodical, the *Philosophical transactions* being used for the longer papers.

095. *L'institut: Journal des académies et sociétés scientifiques de la France et de l'étranger.* Paris, 1833+. Size: 1833-63, 3. Ceased 1876.
Scudder 1395. Bolton 2195.

096. Congrès Scientifique de France. (Annual meetings in different French cities, beginning in 1833)
Sessions (Procès-verbaux et mémoires). 1833+. Size: 1833-61, 2; 1864-73, 1. Ceased 1879. Cf. entry 0218.
Scudder 998.

097. K. Bayerische Akademie der Wissenschaften. München. See also 063.
Gelehrte Anzeigen. 1835+. Size: 1835-60, 3. Cont'd from 1860 as *Sitzungsberichte.* Size: 1860-73, 2; 1901-05, 4.
Scudder 3109.

098. Académie des Sciences. Paris. See also 066.
Comptes rendus hebdomadaires des séances. 1835+. Size: 1835-63, 13; 1864-73, 17; 1901-05, 20 (chemistry 28%, physics 15%, physiology 14%, botany 9%, mathematics 7%, zoology 6%, astronomy 4%, geology 4%, other 13%)
See Index. Scudder 1397.

099. Naturforschende Gesellschaft in Basel.
Bericht über die Verhandlungen. 1835+. Size: 1835-52, 2. Cont'd from 1854 as *Verhandlungen.* Size: 1901-05, 1.
Scudder 2124.

0100. Leicester Literary and Philosophical Society.
Transactions. 1835+. Size: 1901-05, 1.

0101. Académie Impériale des Sciences de St. Pétersbourg. See also 055.
Bulletin (scientifique). 1836+. Size: 1836-58, 5; 1860-73, 4; 1901-05, 4.
Scudder 3707. Documentation: Besterman, 1945. ("This most complicated of all periodicals....")

0102. Royal Irish Academy. Dublin. See also 043.
Proceedings. 1836 (1841)+. Size: 1841-64, 3. Cont'd past 1900.
Scudder 93.

0103. K. Preussische Akademie der Wissenschaften. Berlin. See also 056.
Bericht über die zur Bekanntmachung geigneten Verhandlungen. 1836+. Size: 1836-55, 4. Cont'd from 1856 as *Monatsberichte.* Size: 1856-73, 9. Cont'd from 1882 as *Sitzungsberichte.* Size: 1901-05, 11 (chemistry 27%, mineralogy 15%, physics 13%, mathematics 13%, geology 8%, mechanics 6%, astronomy 6%, other 12%)
Scudder 2370.

0104. Physikalischer Verein. Frankfurt a.M. (Founded 1824)
Jahresbericht. 1837 (1838)+. Size: 1858-60, 5; 1901-05, 2 (various sciences).
Scudder 2605.

0105. K. Gesellschaft der Wissenschaften. Göttingen. See also 020 and 0124.
Abhandlungen. (Successor to its *Commentationes recentiores*--020) 1838/41 (1843)+. Size: 1838-63, 1; 1864-73, 1; 1901-05, 1.
Scudder 2678.

0106. American Philosophical Society. Philadelphia. See also 029.
(a) *Proceedings.* 1838 (1840)+. Size: 1840-63, 1; 1864-73, 4; 1901-05, 2.
Scudder 4240.
(b) ——— *Early proceedings ... compiled ... from the manuscript minutes of its meetings from 1744 to 1838.* Philadelphia, 1884.

0107. Skandinaviska Naturforskare och Läkare. [Scandinavian Scientists and Physicians] A *Wandergeschellschaft* which held annual meetings in different Scandinavian cities, beginning in 1839.
Fördhandlingar ... Möte. 1839 (1840)+. Size: 1839-56, 3. Ceased 1929?
Scudder 571.

0108. Riunione (*or* Congresso) degli Scienziati Italiani. (Annual meetings in different Italian cities, beginning in 1839)
Atti. 1839 (1840)+. Size: 1840-47, 7. Ceased 1847.
Scudder 1784.
For political reasons the association was suppressed after the 1847 meeting, though some later meetings were held (in 1862, 1873, and 1875). The Società Italiana per il Progresso delle Scienze, formed in 1907, regards itself as the continuation of the original association.

0109. Societas Scientiarum Fennica. Helsingfors. See also 0152.
Commentationes. 1840 (1842)+. Cont'd from 1847 as *Acta.* Size: 1840-63, 2; 1901-05, 1.
Scudder 3666.

0110. Magyar Orvosok és Természetvizsgálók Nagygyülése. [Congress of Hungarian Physicians and Naturalists]
Munkálatok. [*Transactions*] 1840? (1841)+. Ceased 1906?
Scudder 3346.

0111. R. Istituto Lombardo di Scienze (Lettere) ed Arti. Milano. See also 0185.
Giornale (successor to 068). 1841+. Size: 1841-56, 1. Cont'd

from 1858 as *Atti*. Size: 1856-62, 1. Ceased 1862. Presumably 0185 is its successor.
Scudder 1898.

0112. (Imp. Reg.) Istituto Veneto di Scienze, Lettere ed Arti. Venezia.
Atti. 1841+. Size: 1841-64, 3; 1865-73, 4; 1901-05, 4.
Scudder 2096.

0113. *Annali di fisica, chimica e matematiche*. Milano, 1841-47.
Edited by G.A. Majocchi. Size: 1841-47, 11.
Scudder 1879. Bolton 404.
Included despite its short duration because of its remarkable size. Its termination was apparently due to political reasons as (according to Poggendorff) Majocchi had to leave Milan in consequence of the revolution of 1848.

0114. (Royal) Philosophical Society of Glasgow.
Proceedings. 1841/42 (1842)+. Size: 1841-64, 1; 1865-73, 2; 1901-05, 1.
Scudder 141.

0115. R. Accademia delle Scienze e Belle Lettere di Napoli. (Sezione della Società Real Borbonicà) See also 070 and 0174.
Rendiconto. 1842+. Size: 1842-50, 4. For continuation after 1860 see 0175.
Scudder 1934.

0116. Naturforschende Gesellschaft in Bern.
Mitteilungen. 1843 (1844)+. Size: 1843-63, 2; 1864-73, 4; 1901-05, 1.
Scudder 2140.

0117. Société Royale des Sciences de Liège.
Mémoires. 1843+. Size: 1843-63, 1. Cont'd past 1900.
Scudder 984.

0118. K. Vetenskaps-Akademien. Stockholm. See also 074 and 0216.
Öfersigt af ... fördhandlingar. 1844+. Size: 1844-63, 4; 1864-73, 3; 1901-05, 3.
Scudder 693x.

0119. Verein für Naturkunde im Herzogthum Nassau. Wiesbaden. (Founded 1829) From 1864 Nassauischer Verein für Naturkunde.
Jahrbücher. 1844+. Size: 1901-05, 2.
Scudder 3321.

0120. Royal Society of Edinburgh. See also 044.
Proceedings. 1832/44 (1845)+. Size: 1845-63, 2; 1864-73, 4; 1901-05, 5.
Scudder 124.

0121. Académie des Sciences, Belles-Lettres et Arts de Lyon. (Founded 1700; in the 18th century it published only one volume of *Mémoires couronnés*)
Mémoires. 1845+. Size: 1851-63, 2; 1864-73, 1. Cont'd past 1900.
Scudder 1176.

0122. *Scientific American*. New York, 1845+. (From 1876 accompanied by *Scientific American Supplement*) Size: 1901-05, 7 (incl. Supplement).
Scudder 4218. Bolton 4274 & p. 943.

0123. Naturforschender Verein zu Riga.
Correspondenzblatt. 1845+. Size: 1846-63, 2; 1864-73, 1. Cont'd past 1900.
Scudder 3702.

0124. K. Gesellschaft der Wissenschaften. Göttingen. See also 0105.
Nachrichten. 1845+. Size: 1845-62, 2; 1863-73, 5; 1901-05, 4.
Scudder 2678, 2677.

0125. Verein für Vaterländische Naturkunde in Württemberg. Stuttgart.
Jahreshefte. 1845+. Size: 1845-58, 2; 1864-73, 2; 1901-05, 4.
Scudder 3268.

0126. K. Sächsische Gesellschaft der Wissenschaften. Leipzig. See also 0139.
Bericht über die Verhandlungen. 1846 (1848)+. Size: 1846-63, 2; 1864-73, 2; 1901-05, 3.
Scudder 3004.

0127. American Academy of Arts and Sciences. Boston. See also 041.
Proceedings. 1846 (1848)+. Size: 1852-63, 1; 1864-73, 1; 1901-05, 4.
Scudder 3974.

0128. Smithsonian Institution. Washington. (Founded 1846) See also 0137 and 0176.
(Annual) Report. 1846 (1847)+. Size: 1850-63, 1; 1864-73, 2; 1901-05, 4.
Scudder 4350a.

0129. *(Supplément à la) Bibliothèque universelle. Archives des sciences physiques et naturelles*. (Continuation of 048c) Genève, 1846+. Edited by A.A. de La Rive, J.C. Marignac, J. Pictet, et al. Size: 1846-57, 3; 1858-73, 4. Cont'd from 1878 as *Archives des sciences physiques et naturelles*. Size: 1901-05, 11 (chemistry 31%, zoology 20%, physics 12%, meteorology 10%, geology 9%, botany 7%, geography 7%, other 4%).
See Index. Scudder 2154. Bolton 933E.

0130. Naturforschende Gesellschaft in Zürich. (Founded 1746; publication record dates from 1761)
Mittheilungen. 1846 (1847)+. Size: 1847-56; 2. Cont'd from 1856 as *Vierteljahrsschrift*. Size: 1856-63, 3; 1864-73, 2; 1901-05, 1.
Scudder 2206.

0131. Académie des Sciences et Lettres de Montpellier. (Founded 1706; fragmentary publication record in the 18th century; suppressed in 1793 and re-founded in 1795 as Société Libre des Sciences et Belles-Lettres; suppressed again in 1815 and re-founded again in 1846)
Mémoires de la Section des sciences. 1847+. Size: 1847-60, 3; 1864-73, 3. Ceased 1912.
Scudder 1228 (also 1240 and 1241).

0132. Accademia Pontificia de' Nuovi Lincei. Roma. (Founded in 1847 by Pope Pius IX as a revival of the 17th-century Accademia de' Lincei. In 1870 it split into two separate academies; see 0211.)
Atti. 1847/48 (1851)+. Size: 1847-63, 2; 1864-73, 2; 1901-05, 1.
Scudder 2020.

0133. Oberhessische Gesellschaft für Natur- und Heilkunde. Giessen. (Founded 1833; re-organised 1846)
Bericht. 1847+. Size: 1847-63, 2. Ceased 1905.
Scudder 2642.

0134. *Corrispondenza scientifica di Roma*. 1848+ Edited by E. (Fabbri) Scarpellini. Size: 1848-63, 2. Ceased 1875?
Scudder 2030. Bolton 1363.

0135. K. Akademie der Wissenschaften. Wien. (Founded 1847) See also 0141, 0146, and 0184.
Sitszungsberichte (Math.-naturwiss. Classe). 1848+. Size: 1848-63, 11; 1864-73, 11; 1901-05, 11 (chemistry 33%, physics 22%, meteorology 10%, mathematics 9%, geology 8%, botany 6%, other 12%).
See Index. Scudder 3560.

0136. American Association for the Advancement of Science. (Founded 1848; for its predecessor see 0470. Holds annual meetings in different American cities.)
Proceedings. 1848+. (Suspended 1860-66) Size: 1848-60, 2; 1866-73, 6. From 1901 the *Proceedings* were published in *Science* (0244).
Scudder 3937.

0137. Smithsonian Institution. Washington. See also 0128 and 0176.
Smithsonian contributions to knowledge. 1848+. Size: 1848-63, 1; 1864-73, 1; 1901-05, 1.
Scudder 4350e. Bolton 4342.

0138. Royal Society of Van Diemen's Land (*from 1856* of Tasmania). Hobart Town.
Papers and proceedings. 1849 (1851)+. Size: 1863-73, 1. Cont'd past 1900.
Scudder 3834, 3833.

0139. K. Sächsische Gesellschaft der Wissenschaften. Leipzig. See also 0126.
Abhandlungen. 1850 (1852)+. Size: 1852-63, 1; 1901-05, 1.
Scudder 3004.

0140. Accademia delle Scienze dell'Istituto di Bologna. See also 090.
Memorie. (Successor to 013) 1850+. Size: 1850-61, 2; 1862-73, 2; 1901-05, 2.
Scudder 1798.

0141. K. Akademie der Wissenschaften. Wien. See also 0135.
Denkschriften (Math.-naturwiss. Classe). 1850+. Size: 1850-64, 2; 1901-05, 2. Contains bigger papers than the *Sitzungsberichte* (0135).
Scudder 3560.

0142. R. Academia de Ciencias (Exactas, Físicas y Naturales). Madrid.
Memorias. 1850+. Size: 1850-63, 1. Cont'd past 1900.
Scudder 1726.

0143. Physikalisch-Medicinische Gesellschaft in Würzburg. (Founded 1849) See also 0236.
Verhandlungen. 1850+. Size: 1850-60, 2. Cont'd from 1860 as *Würzburger naturwissenschaftliche Zeitschrift*. Size: 1860-67,

6. Cont'd from 1868 (1869) as *Verhandlungen*. Size: 1901-05, 2.
Scudder 2572, 3330a.

0144. *Revista de los progresos de las ciencias exactas, fisicas y naturales*. Madrid, 1850+. Size: 1850-62, 1. Ceased 1886.
Scudder 1758. Bolton 3999.

0145. Royal Institution of Great Britain. See also 067.
Notices of the proceedings, later *Proceedings*. 1851 (1854)+. Size: 1851-63, 4; 1864-73, 2; 1901-05, 1.
Scudder 309.

0146. K. Akademie der Wissenschaften. Wien. See also 0135.
Almanach. 1851+. Size: 1864-73, 1. Cont'd past 1900.
Scudder 3560.

0147. *Die Natur: Zeitung zur Verbreitung naturwissenschaftlicher Kenntniss und Naturanschauung. Für Leser aller Stände*. Halle, 1852+. From 1875 published by the Deutscher Humboldt-Verein. Size: 1901, 11. Ceased 1901.
Scudder 2726, 2711. Bolton 3227 & p. 854.

0148. Société des Sciences Naturelles (*from 1879* et Mathématiques) de Cherbourg.
Mémoires. 1852 (1853)+. Size: 1852-63, 2; 1864-73, 1; 1901-05,1.
Scudder 1128.

0149. *Cosmos: Revue encyclopédique hebdomadaire des progrès des sciences* (later *et de leurs applications aux arts et à l'industrie*). Paris, 1852+. Founded by B.R. de Monfort and edited until 1863 by F. Moigno who then ceased his connection with it and established the rival journal *Les mondes* (0179). Size: 1852-63, 1; 1864-70, 3. Ceased 1870.
Scudder 1371. Bolton 1368.

0150. Naturforschender Verein (*later* Gesellschaft). Bamberg. (Founded 1834)
Bericht. 1852+. Size: 1852-61, 1. Cont'd past 1900.
Scudder 2255.

0151. Dorpater Naturforscher-Gesellschaft. *Later* Naturforscher Gesellschaft bei der Universität Jurjev. (Dorpat was later called Jurjev)
Sitzungsberichte. 1853+. Size: 1901-05, 1.
Scudder 3655c.

0152. Finska Vetenskaps-Societeten. Helsingfors. (*Formerly* Societas Scientiarum Fennica; see 0109)
Öfversigt af förhandlingar. 1838/? (1853)+. Size: 1853-74, 2; 1901-05, 2. Ceased 1922.
Scudder 3662.

0153. Naturforschende Gesellschaft zu Halle. (Founded 1779; fragmentary publication record from 1783)
(a) *Abhandlungen*. 1853 (1854)+. Size: 1856-62, 1; 1901-05, 1. Ceased 1919.
(b) *Bericht über die Sitzungen*. 1853+. Published with the *Abhandlungen* until 1880, then separately. Size: 1853-62, 1, 1863-73, 2. Ceased 1892.
Scudder 2493, 2727.

0154. *Zeitschrift für (die gesammten) Naturwissenschaften*. Halle, 1853+. Published by the Naturwissenschaftlicher Verein in Halle (founded 1848). Size: 1853-63, 4; 1864-73, 11; 1901-05, 3.
Scudder 2729, 2403. Bolton 4903 & p. 999.

0155. K. Akademie van Wetenschappen. Amsterdam. (The origins of the academy go back to 1808, if not earlier, but it was re-organised in 1853.)
Verslagen en mededeelingen. 1853+. Size: 1853-64, 2; 1865-73, 3. For continuation after 1891 see 0266.
Scudder 720.

0156. Philosophical Society of Adelaide. *From 1879* Royal Society of South Australia.
Annual report and transactions. 1853 (1854)+. Size: 1865-70, 1. Ceased 1872. Succeeded in 1877 (1878) by *Transactions, proceedings and report*. Size; 1901-05, 1.
Scudder 3827.

0157. Naturforschende Gesellschaft von Graubünden. Chur.
Jahresbericht. 1854/55 (1856)+. Size: 1854-63, 1; 1864-73, 1; 1901-05, 1.
Scudder 2145.

0158. Niederrheinische Gesellschaft für Natur- und Heilkunde. Bonn. (Founded 1818; re-organised 1839)
Sitzungsberichte. 1854+. Published with 0621a until 1868, then separately. Size: 1864-73, 9; 1901-05, 3.
Scudder 2412.

0159. California Academy of (Natural) Sciences. San Francisco.
Proceedings. 1854+. Size: 1860-73, 2. Cont'd past 1900.
Scudder 4306.

0160. Société des Sciences Physiques et Naturelles de Bordeaux.
Mémoires. 1854 (1855)+. Size: 1854-63, 1; 1864-73, 2; 1901-05,1.
Scudder 1075.

0161. Philosophical Society (*from 1860* Royal Society) of Victoria. Melbourne.
Transactions. 1854 (1855)+. Size: 1860-65, 1; 1866-73, 2; 1901-05, 3.
Scudder 3847, 3846, 3848.

0162. Naturforschende Gesellschaft. Freiburg i.B. (Founded 1821; re-organised 1846)
Berichte über die Verhandlungen. 1855 (1858)+. Size: 1858-62, 4. Cont'd from 1886 as *Berichte*. Size: 1901-05, 2.
Scudder 2630.

0163. Royal Dublin Society. See also 051.
Journal. 1856 (1858)+. Size: 1856-63, 1; 1864-73, 2. Cont'd from 1877 as *Scientific proceedings* and *Scientific transactions*. Size: 1901-05, 2 (both together).
Scudder 86.

0164. Academy of Science of St. Louis.
Transactions. 1856 (1857)+. Size: 1856-63, 2; 1901-05, 2.
Scudder 4295.

0165. Essex Institute. Salem, Mass.
Proceedings. 1848 (1856)+. Size: 1856-64, 1. Cont'd from 1869 (1870) as *Bulletin*. Size: 1869-73, 2. Ceased 1898.
Scudder 4302.

0166. Naturhistorisch-Medicinischer Verein. Heidelberg. (Founded 1856)
Verhandlungen. 1857+. Size: 1857-60, 3; 1901-05, 1.
Scudder 2794.

0167. Literary and Philosophical Society of Manchester. See also 039.
Proceedings. 1857 (1860)+. Size: 1857-73, 2; 1901-05, 1.
Scudder 494.

0168. Videnskabs-Selskabet i Christiania.
Forhandlinger. 1858 (1859)+. Size: 1858-63, 1; 1901-05, 1.
Scudder 597.

0169. St. Gallische Naturwissenschaftliche Gesellschaft. St. Gallen.
Bericht. 1858 (1860)+. Size: 1860-63, 2; 1864-73, 1; 1901-05, 1.
Scudder 2178.

0170. K. Böhmische Gesellschaft der Wissenschaften. Prag. See also 040.
Sitzungsberichte. 1859+. Size: 1859-63, 1; 1864-73, 2; 1901-05, 2.
Scudder 3492.

0171. Verein zur Verbreitung Naturwissenschaftlicher Kenntnisse. Wien.
Schriften. 1860/61 (1862)+. Size: 1860-69, 2; 1901-05, 1.
Scudder 3639.

0172. K. Physikalisch-Ökonomische Gesellschaft zu Königsberg. (Founded 1789; fragmentary publication record from 1792)
Schriften. 1860 (1861)+. Size: 1860-73, 1; 1901-05, 2.
Scudder 2869.

0173. Naturwissenschaftliche Gesellschaft 'Isis' in Dresden. (Founded 1833) See also 0622.
Sitzungsberichte. 1861+. Size: 1861-73, 2; 1901-05, 2.
Scudder 2517.

0174. R. Accademia delle Scienze Fisiche e Matematiche. (Sezione della Società Reale di Napoli) Before 1861 its name was R. Accademia delle Scienze e Belle Lettere di Napoli: see 0115. See also 0175.
Atti. (Successor to 070) 1861 (1863)+. Size: 1863-73, 1; 1901-05, 1.
Scudder 1959.

0175. R. Accademia delle Scienze Fisiche e Matematiche. Napoli. See also 0174.
Rendiconto. (Continuation of 0115) 1862+. Size: 1862-73, 8; 1901-05, 3.
Scudder 1959.

0176. Smithsonian Institution. Washington. See also 0128 and 0137.
Smithsonian miscellaneous collections. 1856/? (1862)+. Size: 1862-73, 1; 1901-05, 3.
Scudder 4350. Bolton 4344.

0177. Devonshire Association for the Advancement of Science, Literature, and Art. Plymouth.
Report and transactions. 1862 (1863)+. Size: 1862-73, 2; 1901-05, 1.
Scudder 524.

0178. Naturwissenschaftlicher Verein. Karlsruhe. (Founded 1840; reorganised 1862)
Verhandlungen. 1862 (1864)+. Size: 1901-05, 1.
Scudder 2845.

0179. *Les mondes. Revue hebdomadaire des sciences et de leurs applications aux arts et à l'industrie*. Paris, 1863+. Edited by F. Moigno (see 0149). Size: 1863-73, 8. Cont'd from 1881 as *Cosmos. Les mondes. Revue....* Cont'd from 1885 as *Le cosmos. Revue....* Size : 1901-05, 4.
Scudder 1452. Bolton 3122 & p. 848.

0180. *Revue des cours scientifiques de la France et de l'étranger*. Paris, 1863+. Edited by O. Barot et al. Cont'd from 1871 as *La revue scientifique de la France et de l'étranger. Revue des cours scientifiques*. Size: 1869-73, 2. Cont'd from 1884 as *Revue scientifique* (also known as *Revue rose*). Size: 1901-05, 4.
Scudder 1516. Bolton 4082 & p. 923.

0181. Naturwissenschaftlicher Verein für Steiermark. Graz.
Mittheilungen. 1863+. Size: 1863-71, 2; 1901-05, 2.
Scudder 3408.

0182. National Academy of Sciences. Washington. (Founded 1863. An honorary body which was not of much importance until the 20th century.)
Proceedings. 1863+. Size: 1901-05, 1.
Scudder 4335.

0183. *Quarterly journal of science*. London, 1864+. (Successor to 071) Edited by J. Samuelson et al., and from 1871 by W. Crookes. Size: 1864-73, 1. Cont'd from 1880 as *Journal of science*. Ceased 1885.
Scudder 421. Bolton 3827.

0184. K. Akademie der Wissenschaften. Wien. See also 0135.
Anzeiger (Math.-naturwiss. Classe). 1864+. Size: 1864-73, 1; 1901-05, 1.
Scudder 3560.

0185. R. Istituto Lombardo di Scienze e Lettere. Milano. See also 0111.
Rendiconti (Classe di Sci. mat. e nat.). (Presumably the successor to 0111) 1864+. Size: 1864-73, 5; 1901-05, 5.
Scudder 1898.

0186. Ecole Normale Supérieure. Paris.
Annales scientifiques. 1864+. Founded by L. Pasteur. Size: 1864-73, 1; 1901 05, 1 (all mathematics).
Scudder 1378. Bolton 341 & p. 625.

0187. *Jenaische Zeitschrift für Medicin und Naturwissenschaft* (from 1874 *für Naturwissenschaften*). 1864+. Published by the

Medicinisch-naturwissenschaftliche Gesellschaft zu Jena (founded 1853). Size: 1864-73, 2; 1901-05, 3.
Scudder 2828. Bolton 2313.

0188. Lund. Universitet.
Årsskrift. 1864+. Size: 1864-73, 1; 1901-05, 1.
Scudder 675.

0189. R. Accademia delle Scienze di Torino. See also 065.
Atti. 1865/66 (1866)+. Size: 1865-73, 3; 1901-05, 5.
Scudder 2053.

0190. *Gaea: Natur und Leben. Zeitschrift zur Verbreitung und Hebung naturwissenschaftlicher ... Kenntnisse.* Köln, 1865+. Edited by H.J. Klein. Size: 1865-68, 1; 1901-05, 2.
Scudder 2481. Bolton 1797.

0191. Physikalisch-Medicinische Societät. Erlangen. (Founded 1808; fragmentary publication record before 1865)
Verhandlungen. 1865 (1867)+. Cont'd from 1870 (1871) as *Sitzungsberichte.* Size: 1871-74, 1; 1901-05, 3.
Scudder 2604, 2573.

0192. *Hardwicke's science-gossip.* London, 1865+. Cont'd from 1894 as *Science-gossip.* Size: 1901, 11 (chiefly botany and zoology). Ceased 1902.
Scudder 299. Bolton 2038 & p. 762.

0193. Cambridge Philosophical Society. See also 077.
Proceedings. 1843/63 (1865)+. Size: 1866-76, 2; 1901-05, 5.
Scudder 61.

0194. *The English mechanic (and mirror of science).* London, 1865+. Size: 1901-05, 4 (all astronomy).
Scudder 268. Bolton 1578.

0195. Philosophical Society (*from 1866* Royal Society) of New South Wales. Sydney. (Founded 1850)
Transactions. 1862/65 (1866)+. Size: 1867-73, 1. Cont'd from 1876 as *Journal and proceedings.* Size: 1901-05, 1.
Scudder 3858, 3859.

0196. Teyler's Stichting. (Musée Teyler) Haarlem.
Archives du Musée Teyler. 1866+. Size: 1868-74, 1; 1901-05, 1 (mathematics and physical sciences).
Scudder 853. Bolton 641 & p. 646.

0197. *Archives néerlandaises des sciences exactes et naturelles.* La Haye/Harlem, 1866+. Published by the Société Hollandaise des Sciences à Harlem (Hollandse Maatschappij der Wetenschappen) Size: 1866-73, 2; 1901-05, 5.
Scudder 857a. Bolton 627.

0198. Gesellschaft zur Beförderung der gesammten Naturwissenschaften. Marburg. (Founded 1816)
Sitzungsberichte. 1866+. Size: 1901-05, 4.
Scudder 3080.

0199. Towarzystwo Naukowe Krakowskie. [Cracow Society of Science] (Publications date from 1817. In 1873 it became the Cracow Academy of Science: see 0224)

Sprawozdanie komisyi fizyograficznéj. [Report of the physiographical commission] 1866 (1867)+. Size: 1901-05, 2 (chiefly botany, meteorology, and zoology).
Scudder 4445c. Bolton 8048.

0200. Società dei Naturalisti (*from 1899* e Matematici) in Modena.
Annuario. 1866+. Size: 1866-73, 1. Cont'd from 1883 as *Atti*. Size: 1901-05, 1.
Scudder 1925.

0201. Siezd Russkikh Estestvoispuitatelei (*later* i Vrachei). [Meeting of Russian Naturalists (and Physicians)] Meetings held every few years in different cities of the Russian Empire, beginning in 1867/68.
Trudui [*Transactions*]. 1867 (1868)+. Cont'd later as *Dnevnik XI* [*Journal XI*]. Size: 1901-05, 5.
Scudder 3653.

0202. Naturwissenschaftlicher Verein. Bremen. (Founded 1864)
Abhandlungen. 1864/66 (1868)+. Size: 1868-73, 1; 1901-05, 2.
Scudder 2433.

0203. *Jornal de sciencias mathematicas, physicas e naturaes*. Lisboa, 1868+. Published by the Academia Real das Sciencias de Lisboa. Size: 1868-71, 2. Ceased 1927.
Scudder 1720f. Bolton 2323.

0204. Kansas Natural History Society. From 1872 Kansas Academy of Science.
Transactions. 1868 (1872)+. Size: 1901-05, 2.
Scudder 4320.

0205. New Zealand Institute. Wellington.
Transactions and proceedings. 1868 (1869)+. Size: 1868-74, 8; 1901-05, 5.
Scudder 3863.

0206. *Nature. A weekly illustrated journal of science*. (Sub-title varies) London, 1869+. Edited by J.N. Lockyer until 1919. Size: 1870-74, 6; 1901-05, 11.
See Index. Scudder 392. Bolton 3259. Documentation: Fifield, 1969; Gregory, 1943; MacLeod, 1969; *Nature*, 1969.

0207. *Természettudományi közlöny* [*Journal of natural science*] Budapest, 1869+. Published by the Magyar Természettudományi Társulat [Hungarian Society of Natural Science]. Size: 1901-05, 4.
Scudder 3371. Bolton 4472.

0208. Naturwissenschaftlicher Verein von Neu-Vorpommern und Rügen. Greifswald. (Founded 1866)
Mittheilungen. 1869+. Size: 1901-05, 1.
Scudder 2357.

0209. Wisconsin Academy of Sciences, Arts and Letters. Madison.
Transactions. 1870 (1872)+. Size: 1870-74, 1; 1901-05, 2.
Scudder 4128.

0210. Kievskoe Obshchestvo Estestvoispuitatelei. (Société des Naturalistes de Kiev)
Zapiski. *(Mémoires)*. 1870+. Size: 1901-05, 1.
Scudder 3680.

0211. R. Accademia dei Lincei. Roma. (Formed in 1870 by division--resulting from political events--of the Accademia Pontificia de' Nuovi Lincei (0132). It became the national academy of Italy.)
Atti. 1870+. Size: 1870-73, 3. Cont'd from 1892 as *Rendiconti*. Size: 1901-05, 11 (chemistry 20%, geology 17%, mathematics 15%, zoology 9%, physics 9%, physiology 7%, botany 6%, astronomy 6%, other 11%).
See Index. Scudder 2019 & errata.

0212. Varshavskia Imp. Universitetskia. (Université Impériale de Varsovie) (The Russian university in Warsaw, founded in 1869 --as distinct from the Polish university, founded in 1816.)
Izvestiya. (Bulletin) 1870+. Size: 1901-05, 2.

0213. Naturwissenschaftlich-Medicinischer Verein. Innsbruck.
Berichte. 1870+. Size: 1870-73, 3. Cont'd past 1900.
Scudder 3426.

0214. Imp. Sankt-Peterburgskoe Obshchestvo Estestvoispuitatelei. (Société Impériale des Naturalistes de St. Pétersbourg)
Trudui. (Travaux) 1870+. Size: 1901-05, 6.
Scudder 3745.

0215. Obshchestvo Estestvoispuitatelei pri Imp. Kazanskom Universitetye. [Society of Naturalists at the Imp. University of Kazan]
Trudui. [Transactions] 1871+. Size: 1901-05, 1.
Scudder 3675.

0216. K. Vetenskaps-Akademien. Stockholm. See also 0118.
Bihang till handlingar. (Supplement to 014a) 1872 (1873)+. Size: 1901-05, 3.
Scudder 693u.

0217. *Popular science monthly*. New York, 1872+. Size: 1901-05, 6.
Scudder 4216. Bolton 3706 & p. 894.

0218. Association Française pour l'Avancement des Sciences. (Held annual meetings in different French cities, beginning in 1872) Cf. entry 096.
Comptes rendus. 1872 (1873)+. Size: 1872-73, 11; 1901-05, 10 (zoology 26%, botany 14%, physics 14%, meteorology 10%, geology 9%, physiology 6%, other 21%).
Scudder 993.

0219. Novorossiiskoe Obshchestvo Estestvoispuitatelei. (Société des Naturalistes de la Nouvelle Russie) Odessa.
Zapiski. (Mémoires) 1872+. Size: 1901-05, 1.
Scudder 3694.

0220. *La nature. Revue des sciences et de leurs applications aux arts et à l'industrie*. Paris, 1873+. Edited by G. Tissandier. Size: 1901-05, 10.
Scudder 1467. Bolton 3260 & p. 856.

0221. Naturwissenschaftlicher Verein für Schleswig-Holstein. Kiel. (Founded 1872)
Schriften. 1873 (1875)+. Size: 1901-05, 1.
Scudder 2853.

0222. Société des Sciences de Nancy. (Continuation after 1870 of 0607)
Bulletin. 1873 (1874)+. Size: 1901-05, 1.
Scudder 1247.

0223. Philosophical Society of Washington.
Bulletin. 1871/74 (1874)+. Size: 1901-05, 1. Ceased 1910.
Scudder 4346.

0224. Akademia Umiejetności w Krakowie. [Academy of Sciences in Cracow] (For predecessor see 0199) See also 0259.
Rozprawy. [*Proceedings*] 1874+. Size: 1901-05, 7.

0225. Naturforschende Gesellschaft zu Leipzig. (Founded 1818)
Sitzungsberichte. 1874+. Size: 1901-05, 1.
Scudder 2975.

0226. Naturwissenschaftlicher Verein in Hamburg. (Founded 1837)
Verhandlungen. 1875 (1877)+. Size: 1901-05, 2.
Scudder 2763.

0227. *Kosmos. Czasopismo Polskiego Towarzystwa Przyrodnikow imienia Kopernika*. [*Cosmos. Journal of the Polish Society of Naturalists founded in honour of Copernicus*] Lwów, 1875+. Size: 1901-05, 3.
Bolton 2575.

0228. *Archiv for mathematik og naturvidenskab*. Kristiania, 1876+. Edited by S. Lie et al. Size: 1901-05, 1 (various sciences).
Scudder 576. Bolton 576.

0229. *Naturen. Illustret månedstidsskrift for populaer naturvitenskap*. Bergen, 1877+. Published by the University of Bergen. Size: 1901-05, 2.
Bolton 3261.

0230. New York Academy of Sciences. (For its predecessor see 0599)
Annals. 1877+. Size: 1901-05, 1.

0231. South African Philosophical Society. Cape Town.
Transactions. 1877 (1878)+. Size: 1901-05, 2.

0232. Imperial University of Tokyo.
Memoirs of the Science Department. 1879+. Cont'd from 1886 (1887) as *Journal of the College of Science*. Size: 1901-05, 2.
Bolton 6792.

0233. Verein für Naturwissenschaft zu Braunschweig. (Founded 1862)
Jahresbericht. 1879/80 (1880)+. Size: 1901-05, 1.

0234. Johns Hopkins University. Baltimore, Md.
Circulars. 1879+. Size: 1901-05, 1.

0235. *Knowledge*. London, 1881+. Edited by R.A. Proctor. Size: 1901-05, 3 (chiefly astronomy and meteorology).
Bolton 2569.

0236. Physikalisch-Medicinische Gesellschaft in Würzburg. See also 0143.
Sitzungsberichte. 1881+. Size: 1901-05, 3.

0237. Royal Society of Canada.
Proceedings and transactions. Montreal, etc., 1882 (1883)+. Size: 1901-05, 2.

0238. Magyar Tudományos Akademia. (Ungarische Akademie der Wissenschaften) Budapest. (The Academy's publication record dates from 1778; see Scudder 3372)

Mathematikai és természettudományi értestitö. (Mathematischer und naturwissenschaftlicher Anzeiger) 1882 (1883)+. Size: 1901-05, 3.
Bolton 6956.

0239. *Mathematische und naturwissenschaftliche Berichte aus Ungarn.* Berlin, Budapest, Leipzig, 1882+. Published with the support of the Magyar Tudományos Akadémia and the K. Magyar Természett-udományi Társulat. Size: 1901-05, 3.
Bolton 6958.

0240. *Wszechświat, tygodnik poświecony naukom przyrodniczyn.* Warszawa, 1882+. Size: 1901-05, 4 (various sciences, mostly biological).

0241. Colorado Scientific Society. Denver.
Proceedings. 1883+. Size: 1901-05, 1.

0242. Elisha Mitchell Scientific Society. (University of North Carolina) Chapel Hill, N.C. (Named after Elisha Mitchell (1793-1857), an early naturalist and geologist of North Carolina)
Journal. 1883 (1884)+. Size: 1901-05, 1.

0243. Naturwissenschaftlicher Verein des Regierungsbezirkes Frankfurt a.d. Oder.
Monatliche Mittheilungen. 1883+. Cont'd from 1891 as *Abhandlungen.* Cont'd from 1892 as *Helios. Abhandlungen....* Size: 1901-05, 1.
Bolton 7106.

0244. *Science.* New York, 1883+. From 1895 edited by J.M. Cattell who in 1900 made it the official organ of the American Association for the Advancement of Science (0136); from 1901 it included the Association's *Proceedings.* Size: 1901-05, 12.
See Index. Bolton 4255 & p. 940. Documentation: DSB--article on Cattell.

0245. Hamburgische Wissenschaftliche Anstalt.
Jahrbuch. 1883+. Size: 1901-05, 1 (most sciences).

0246. Natural Science Association of Staten Island. New Brighton, N.Y.
Proceedings. 1883+. Size: 1901-05, 1.

0247. *Mathematisch-naturwissenschaftliche Mitteilungen.* Tübingen, later Stuttgart, 1884+. Published by the Mathematisch-naturwissenschaftlicher Verein in Württemberg. Size: 1901-05, 1.
Bolton 6957.

0248. *Yearbook of the scientific and learned societies of Great Britain and Ireland.* London, 1884+.
Bolton 8392.

0249. Naturforschende Gesellschaft. Rostock. (Founded 1882)
Sitzungsberichte. 1886+. Size: 1901-05, 2.

0250. *Naturwissenschaftliche Rundschau: Wöchentliche Berichte über die Fortschritte auf dem Gesamtgebiete der Naturwissenschaften.* Braunschweig, 1886+. Size: 1901-05, 5. Cont'd from 1913 as *Die Naturwissenschaften.*
Bolton 7208.

0251. *Naturwissenschaftliche Wochenschrift.* Berlin, 1887+. Size: 1901-05, 9. Ceased 1922.
Bolton 7209.

0252. Accademia Pontificia dei Nuovi Lincei. Roma. See also 0132.
Memorie. 1887+. Size: 1901-05, 1.

0253. Faculté des Sciences (de l'Université) de Toulouse.
Annales. 1887+. Size: 1901-05, 2.
Bolton 5253.

0254. Nederlandsch Natuur- en Geneeskundig Congres. [Dutch Congress for Science and Medicine]
Handelingen. Haarlem, 1887+. Size: 1901-05, 3.

0255. Accademia Gioenia di Scienze Naturali in Catania. See also 085.
Bullettino. 1888+. Size: 1901-05, 1.

0256. Australasian Association for the Advancement of Science. (Holds annual meetings in different cities of Australia and New Zealand, beginning in 1888)
Report. 1888+. Size: 1901-05, 3.

0257. *Himmel und Erde. Illustrierte naturwissenschaftliche Monatsschrift*. Berlin, etc., 1889+. Published by the Gesellschaft Urania A.G. Size: 1901-05, 3 (various sciences)
Bolton 6456.

0258. Princeton College (later University). Princeton, N.J.
Bulletin. 1889+. Size: 1901-04. 1. Ceased 1904.

0259. Akademia Umiejetności w Krakowie. (Académie des Sciences de Cracovie) See also 0224.
Bulletin international. 1889+. Size: 1901-05, 7. The contents are mostly translations into French or German of papers in the Academy's *Rozprawy* (0224).

0260. *Revue générale des sciences pures et appliquées*. Paris, 1890+. Edited by L. Olivier. Size: 1901-05, 7.
Bolton 7751.

0261. Società Ligustica di Scienze Naturali e Geografiche. Genova.
Atti. 1890+. Size: 1901-05, 2.

0262. Česká Akademie Císaře Františka Josefa pro Vědy, Slovesnost a Umění v Praze. [Czech Academy of the Emperor Franz Josef for the Sciences, Letters and Arts in Prague] (Founded 1890)
Rozpravy [*Proceedings*]. 1891+. Size: 1901-05, 3.

0263. Imp. Kharkovskoe Universitet. (Université Impériale de Charkov)
Letopisi (Annales). 1891+. Size: 1901-05, 1.

0264. Indiana Academy of Science. Indianapolis.
Proceedings. 1891 (1892)+. Size: 1901-05, 4.

0265. Fiziko-Matematicheskoe Obshchestvo pri Imp. Kievskom Universitetye. (Société Physico-Mathématique de l'Université Impériale de Kiev)
Otchet (Travaux). 1891+. Size: 1901-05, 1.

0266. K. Akademie van Wetenschappen. Amsterdam. Wis- en Natuurkundige Afdceling. See also 0155.
(a) *Verslagen der zittengen*. (Continuation of 0155) 1892 (1893)+. Size: 1901-05, 11 (chemistry 28%, physics 22%, mathematics 17%, physiology 7%, zoology 5%, geology 3%, other 18%) From 1898 translations of the papers into

English or German were published in (b).
(b) *Proceedings of the Section of Sciences.* 1898 (1899)+.
See Index.

0267. Kansas University. Lawrence.
Kansas university quarterly. 1892+. Size: 1901-05, 1.

0268. Société Scientifique et Médicale de l'Ouest. Rennes.
Bulletin. 1892+. Size: 1901-05, 2.

0269. Ohio State Academy of Science. Columbus.
Annual report. 1892+. Size: 1901-05, 1.

0270. Iowa Academy of Sciences. Des Moines.
Proceedings. 1893+. Size: 1901-05, 3.

0271. Imperialis Universitas Jurjevensis. (Jurjev was known earlier as Dorpat and later as Tartu)
Acta et commentationes. 1893+. Size: 1901-05, 1.

0272. Michigan Academy of Science. Lansing.
Annual report. 1894+. Size: 1901-05, 2.

0273. Deutsche Gesellschaft für Kunst und Wissenschaft in Posen. Naturwissenschaftlicher Verein.
Zeitschrift. 1894+. Size: 1901-05, 2.

0274. Congrès des Sociétés Savantes de Paris et des Départements. Section des Sciences.
Comptes-rendus. 1895+. Size: 1901-05, 3.

0275. University of California. Berkeley.
Chronicle. 1898+. Size: 1901-05, 1.

0276. Washington Academy of Sciences.
Proceedings. 1899+. Size: 1901-05, 2. Included the publications of several affiliated societies in Washington.

MATHEMATICS

..... Ecole Centrale des Travaux Publics. Paris. From 1795 Ecole Polytechnique.
Journal. 1794+. Main entry 046. Chiefly mathematical.

0277. *Annales de mathématiques pures et appliquées.* Nîmes, 1810 +. Edited by J.D. Gergonne. Size: 1811-31, 4. Ceased 1831. From 1836 its place was taken by Liouville's Journal (0280).
Scudder 1259. Bolton 350. Documentation: DSB--article on Gergonne.

0278. *Correspondance mathématique et physique.* Gand/Bruxelles, 1825+. Edited by J.G. Garnier and A. Quetelet. Size: 1825-39, 4. Ceased 1839. Its title was revived in item 0303a.
Scudder 970. Bolton 1350. Documentation: Elkhadem, 1978.

0279. *Journal für die reine und angewandte Mathematik.* Berlin, 1826+. Edited by A.L. Crelle until his death in 1855. (Still known as Crelle's Journal) Size: 1826-63, 4; 1864-73, 3; 1901-05, 3.
Scudder 2341. Bolton 2497. Documentation: DSB--articles on Crelle and Abel.

0280. *Journal de mathématiques pures et appliquées*. Paris, 1836+. Edited by J. Liouville until 1874. Size: 1836-55, 5; 1856-73, 4; 1901-05, 1.
Scudder 1415. Bolton 2374 & p. 794. Documentation: DSB--article on Liouville.

0281. (a) *The Cambridge mathematical journal*. Cambridge, 1839+. Edited by D.F. Gregory and R.L. Ellis. Size: 1839-45, 2. Cont'd as (b).
(b) *The Cambridge and Dublin mathematical journal*. Cambridge, 1846+. Edited by W. Thomson and N.M. Ferrers. Size: 1846-54, 5. Cont'd as (c).
(c) *The quarterly journal of pure and applied mathematics*. London, 1855+. Edited by J.J. Sylvester and N.M. Ferrers. Size: 1855-63, 6; 1863-74, 3; 1901-05, 3.
Scudder 60, 420. Bolton 1193.

0282. *Archiv der Mathematik und Physik*. Greifswald, etc., 1841+. Edited by J.A. Grunert. Size: 1841-63, 6; 1864-73, 5; 1901-05, 5 (nearly all mathematics).
Scudder 2696. Bolton 573 & p. 642.

0283. *Nouvelles annales de mathématiques*. Paris, 1842+. Edited by O. Terquem et al. Size: 1842-61, 8; 1862-73, 5; 1901-05, 5. Ceased 1927.
Scudder 1471. Bolton 349 & p. 626.

0284. *Annali di scienze matematiche e fisiche*. Roma, 1850+. Edited by B. Tortolini. Cont'd from 1858 as *Annali di matematica pura ed applicata*. Size: 1850-60, 7; 1861-73, 2; 1901-05, 1.
Scudder 2022, 2021. Bolton 405.

0285. *Zeitschrift für Mathematik und Physik*. Leipzig, 1856+. Edited by O. Schlömilch and B. Witzschel. Size: 1856-63, 3; 1864-73, 5; 1901-05, 6 (mathematics 47%, mechanics 33%, physics 12%, other 8%). Ceased 1917.
Scudder 3042. Bolton 4862.
Supplement: see 0304.

0286. *Mathematisk tidsskrift*. Kjøbenhavn, 1859+. Edited by H.C.F.C. Schjellerup and C. Tychsen. Cont'd from 1865 as *Tidsskrift for mathematik*. Size: 1865-73, 1. Cont'd from 1890 as *Nyt tidsskrift for matematik*. Size: 1901-05, 1.
Scudder 627. Bolton 2875 (& pp. 829, 965), 8195.

0287. *The Oxford, Cambridge and Dublin messenger of mathematics*. Cambridge, 1862+. Size: 1862-71, 3. Cont'd from 1871 as *The messenger of mathematics*. Size: 1901-05, 4.
Scudder 66. Bolton 3517.

0288. *Giornale di matematiche*. Napoli, 1863+. Edited by G. Battaglini et al. Size: 1863-73, 5; 1901-05, 2.
Scudder 1945. Bolton 1974.

..... Ecole Normale Supérieure. Paris.
Annales scientifiques. 1864+. Main entry 0186. Chiefly mathematical.

0289. London Mathematical Society. (Founded 1865)
Proceedings. 1865+. Size: 1866-73, 3; 1901-05, 4.
Scudder 341.

0290. *Matematicheskii sbornik.* [*Mathematical magazine*] Moskva, 1866+. Published by the Moskovoskoe Matematicheskoe Obshchestvo [Moscow Mathematical Society. (Founded 1864)] Size: 1901-05, 2.
Scudder 3685. Bolton 2851. Documentation: DSB--article on Brashman, N.D.

0291. *Bullettino di bibliografia e di storia delle scienze matematiche e fisiche.* Roma, 1868+. Edited by B. Boncompagni. Largely or entirely historical. Ceased 1887.
Scudder 2025. Bolton 1166. Documentation: DSB--article on Boncompagni.

0292. *Tidskrift för matematik och fysik.* Uppsala, 1868+. Edited by G. Dillner et al. Size: 1868-71, 3. Ceased 1885?
Scudder 714. Bolton 4511.

0293. *Jahrbuch über die Fortschritte der Mathematik.* Berlin, 1868 (1871)+. Edited by C. Ohrtmann and F. Müller. An abstracting periodical. Ceased 1944.
Scudder 2328. Bolton 2268.

0294. *Mathematische Annalen.* Leipzig, 1869+. Edited by A. Clebsch and C. Neumann. Size: 1869-73, 7; 1901-05, 5.
Scudder 2967. Bolton 2869.

0295. *Zeitschrift für mathematischen und naturwissenschaftlichen Unterricht.* Leipzig, 1870+. Edited by J.C.V. Hoffmann. Size: 1901-05, 4 (nearly all mathematics).
Scudder 3043. Bolton 4863.

0296. *Bulletin des sciences mathématiques et astronomiques.* Paris, 1870+. Edited by G. Darboux "sous la direction de la Commission des Hautes Etudes." Chiefly bibliographical. Size: 1870-73, 2. Cont'd from 1877 as *Bulletin des sciences mathématiques.* Size: 1901-05, 1.
Scudder 1350. Bolton 1143 & p. 682.

0297. Société Mathématique de France. Paris. (Founded 1872)
Bulletin. 1872 (1873)+. Size: 1901-05, 4.
Scudder 1589.

0298. *Časopis pro pěstování matematiky a fysiky.* [*Journal for the cultivation of mathematics and physics*] Praha, 1872+. Published by the Jednota Českých Matematatiků a Fysiků [Society of Czech Mathematicians and Physicists]. Size: 1901-05, 1.
Bolton 1218 (cf. 1219 and 618).

0299. Mathematische Gesellschaft in Hamburg. (Founded 1690 as the Kunstrechnungsliebende Societät, a society for popular mathematics)
Mittheilungen. 1873+. Size: 1901-05, 1. The volume for 1890 contains a history of the society from 1690 to 1890.
Documentation: Müller, 1883.

0300. *Periodico di scienze matematiche (e naturali) per l'insegnamento secondario.* Roma, 1873+. Size: 1901-05, 4.
Scudder 2040. Bolton 3556.

0301. (a) *The analyst: A journal of pure and applied mathematics.* Des Moines, Iowa, 1874+. Edited by J.E. Hendricks. Cont'd as (b).

(b) *Annals of mathematics*. Charlottesville, Va., 1884+. Edited by O. Stone and W.H. Thornton. Size: 1901-05, 2.
Scudder 4065. Bolton 287 (& p. 623), 5272.

0302. *Nieuw archief voor wiskunde*. Amsterdam, 1875+. Edited by D. Bierens de Haan. Size: 1901-05, 2.
Bolton 556.

0303. (a) *Nouvelle correspondance mathématique*. Bruxelles, 1876+. Edited by P. Mansion et al. The title was a revival of that of item 0278. Succeeded by (b).
(b) *Mathésis: Recueil mathématique*. Gand, 1881+. Size: 1901-05, 2.
Scudder 980. Bolton 1351, 2876 & p. 829. Documentation: DSB--article on Mansion.

0304. *Abhandlungen zur Geschichte der Mathematik (*later *der mathematischen Wissenschaften)*. Leipzig, 1877+. Edited by M. Curtze and S. Günther. A supplement to *Zeitschrift für Mathematik und Physik* (0285); published independently from 1899. Ceased 1913.

0305. *American journal of mathematics, pure and applied*. Baltimore, 1878+. Edited by J.J. Sylvester. Published under the auspices of the Johns Hopkins University. Size: 1901-05, 2.
Bolton 217 & p. 619.

0306. Kharkovskoe Matematicheskoe Obshchestvo. (Société Mathématique de Kharkov)
Soobshcheniya. (Rapports) 1879+. Size: 1901-05, 1.
Bolton 8026.

0307. *Acta mathematica*. Stockholm, 1882+. Edited by G. Mittag-Leffler. In French and German. Size: 1901-05, 2.
Bolton 5014. Documentation: DSB--article on Mittag-Leffler.

0308. Edinburgh Mathematical Society. (Founded 1883)
Proceedings. 1883+. Size: 1901-05, 1.

0309. *Bibliotheca mathematica: Zeitschrift für Geschichte der Mathematik*. Stockholm, etc., 1884+. Edited by G. Eneström. Size: 1901-05, 3. Ceased 1915.
Bolton 5541.

0310. Circolo Matematico di Palermo. (Founded 1884)
Rendiconti. 1884 (1887)+. Edited by G.B. Guccia. Size: 1901-05, 2.
Documentation: DSB--article on Guccia.

0311. Tokyo Sugaku-Buturigaku Kwai. [Tokyo Mathematico-Physical Society]
Kiji. [*Report*] 1884+. In Japanese and European languages. Size: 1901-05, 2. Several changes of name after 1900.

0312. *Prace matematyczno-fizyczne*. [*Mathematics and physics bulletin*] Warszawa, 1888+. Size: 1901-05, 2 (chiefly mathematics).

0313. Deutsche Mathematiker-Vereinigung. Halle (Founded 1890)
Jahresbericht. Berlin, etc., 1890 (1892)+. Size: 1901-05, 4.

0314. *Monatshefte für Mathematik und Physik*. Wien, 1890+. Edited by G. von Escherich and L. Gegenbauer. Size: 1901-05, 3. The

words "und Physik" were later dropped from the title.
Bolton 7112.

0315. *Revue de mathématiques spéciales*. Paris, 1890+. Edited by B. Niewenglowski. Size: 1901-05, 1.
Bolton 7807.

0316. *Rivista di matematica (Revue de mathématiques)*. Torino, 1891+. Edited by G. Peano. Size: 1901-05, 1. Ceased 1906.
Bolton 7872. Documentation: DSB--article on Peano.

0317. New York (*from 1894* American) Mathematical Society. (Founded 1888)
Bulletin. 1891+. Size: 1901-05, 2.

0318. *Mathematikai és physikai lapok*. [*Mathematical and physical journal*] Budapest, 1892+. Published by the Magyar Tudományos Akadémia [Hungarian Academy of Sciences]. Size: 1901-05, 3 (chiefly mathematics).

0319. *Revue semestrielle des publications mathématiques*. Amsterdam, 1893+. Published under the auspices of the Société Mathématique d'Amsterdam. Ceased 1934.
Bolton 7771.

0320. *American mathematical monthly*. Springfield, Mo., 1894+. Edited by B.F. Finkel and J.M. Colaw. Size: 1901-05, 4.
Bolton 5189.

0321. *Unterrichtsblätter für Mathematik und Naturwissenschaften*. Berlin, 1895+. Published by the Verein zur Förderung des Mathematischen und Naturwissenschaftlichen Unterrichts. Size: 1901-05, 2 (all mathematics).
Bolton 8261.

0322. *Il pitagora. Giornale di matematica*. Palermo, 1895+. Size: 1901-05, 2.

0323. *Wiadomości matematyczne* [*Mathematical sciences*]. Warszawa, 1897+. Size: 1901-05, 2.

0324. International Congress of Mathematicians. [Also in French, German, and Italian] (The first meeting was held at Zurich in 1897)
Proceedings. 1897+.

0325. *Enseignement mathématique*. Paris, 1899+. Published by the Commission Internationale de l'Enseignement Mathématique. Size: 1901-05, 4.

ASTRONOMY

Titles of a large number of periodicals issued by astronomical observatories were found but the great majority of them were not cited in the Royal Society's *Catalogue of Scientific Papers* or the *International Catalogue of Scientific Literature* and so were not included. They mostly contain records of observations or administrative reports.

Astronomical ephemerides and nautical almanacs constitute a special category and are included in an appendix at the end of this section. In the late eighteenth century some of them--notably *Connoissance des Temps* (0351) and Bode's *Astronomisches Jahrbuch* (0356)--played a dual role in that they published astronomical papers and news items as well as ephemerides, and so they can be regarded as the first astronomical periodicals, indeed the first of all the specialized scientific periodicals.

0326. *Monatliche Correspondenz zur Beförderung der Erd- und Himmelskunde*. Gotha, 1800+. Edited by F. von Zach. Primarily astronomical. Size: 1800-13, 4. Ceased 1813. Succeeded by 0327.
Scudder 2669. Bolton 3093. Documentation: Herrmann, 1972.

0327. *Correspondance astronomique, géographique, hydrographique et statistique*. Gênes [i.e. Genoa], 1818+. Edited by F. von Zach. Successor to 0326. Size: 1818-26, 3. Ceased 1826.
Scudder 1855. Bolton 1346. Documentation: Herrmann, 1972.

0328. (Royal) Astronomical Society. London. (Founded 1820) See also 0330.
Memoirs. 1821 (1822)+. Size: 1822-63, 1; 1864-73, 1; 1901-05, 1.
See Index. Scudder 214.

0329. *Astronomische Nachrichten*. Altona, later Kiel, 1823+. Edited by H.C. Schumacher, then from 1854 to 1880 by C.A.F. Peters. Size: 1823-63, 10; 1864-73, 6; 1901-05, 14.
See Index. Scudder 2237. Bolton 712 & p. 649. Documentation: Herrmann, 1972.

0330. (Royal) Astronomical Society. London. See also 0328.
Monthly notices. 1827 (1831)+. Size: 1827-63, 4; 1864-73, 7; 1901-05, 10.
See Index. Scudder 214.

0331. *The astronomical journal*. Cambridge, Mass., 1849+. Edited by B.A. Gould. Suspended between 1861 and 1887. Size: 1851-61, 7; 1901-05, 6.
See Index. Scudder 4016. Bolton 689 & p. 648. Documentation: Herrmann, 1971.

0332. Collegio Romano. Osservatorio. See also 0578.
Memorie. Roma, 1851 (1852)+. Size: 1851-63, 1; 1901-05, 1. Ceased 1921.
Scudder 2039. Bolton 2935.

0333. Harvard College. Astronomical Observatory. (Founded 1847)
Annals. Cambridge, Mass., 1855 (1856)+. Size; 1901-05, 1.
Scudder 4017. Bolton 430.

0334. Astronomische Gesellschaft. Leipzig. (Founded 1863)
Vierteljahrsschrift. 1866+. Size: 1866-73, 1; 1901-05, 2.
See Index. Scudder 2893.

0335. *Sirius. Zeitschrift für populäre Astronomie*. Leipzig, 1868+. Edited by R. Falb. Size: 1901-05, 1.
Scudder 3410. Bolton 4337.

0336. Società degli Spettroscopisti Italiani. (Founded 1871)
Memorie. Palermo, 1872+. Size: 1872-73, 4; 1901-05. 3 (astron-

omy 86%, physics 14%).
See Index. Scudder 1985.

0337. *The observatory. A monthly review of astronomy.* London, 1877 (1878)+. Edited by W.H.M. Christie. Size: 1901-05, 5.
Bolton 3418.

0338. *Ciel et terre. Revue populaire d'astronomie.* Bruxelles, 1880+. Edited by C. Hooreman. Size: 1901-05, 1.
Bolton 1288.

0339. (a) *The sidereal messenger.* Northfield, Minn., 1882+. Edited by W.W. Payne. Cont'd as (b).
(b) *Astronomy and astro-physics.* Ib., 1892+. Edited by W.W. Payne and G.E. Hale. Succeeded by (c).
(c) *The astrophysical journal.* Chicago, 1895+. Edited by G.E. Hale et al. Size: 1901-05, 8.
Bolton 4334 (& p. 946), 5444.

0340. *Bulletin astronomique.* Paris, 1884+. Published by the Observatoire de Paris. Size: 1901-05, 4.
Bolton 5684.

0341. Société Astronomique de France. Paris.
Bulletin. 1887+. Size: 1901-05, 3.

0342. Lick Observatory. (University of California)
Publications. Sacramento, Calif., 1887+. Size: 1901-05, 2.
Bolton 7623.

0343. Astronomical Society of the Pacific. San Francisco.
Publications. 1889+. Size: 1901-05, 2.

0344. *(Klein's) Jahrbuch der Astronomie und Geophysik.* Leipzig, 1890+. Edited by J. Klein. Ceased 1913.
Bolton 6662.

0345. British Astronomical Association. London. (Founded 1890 for amateur astronomers) See also 0348.
Journal. 1890+. Size: 1901-05, 5.
See Index. Documentation: DSB--article on Maunder, E.W.

0346. Vereinigung von Freunden der Astronomie und kosmischen Physik. Berlin. (Founded 1891)
Mitteilungen. 1891+. Size: 1901-05, 1.
Bolton 7083.

0347. Russkoe Astronomicheskoe Obshchestvo. [Russian Astronomical Society] St. Petersburg.
Izvestiya [*Bulletin*]. 1892+. Size: 1901-05, 1.

0348. British Astronomical Association. London. See also 0345.
Memoirs. 1893+. Size: 1901-05, 1.

0349. Leeds Astronomical Society.
Journal and transactions. 1893 (1894)+. Size: 1901-05, 1. Ceased 1922.

0350. *Popular astronomy.* Northfield, Minn., 1893+. Edited by W.W. Payne and C.R. Willard, Goodsell Observatory of Carleton College. Size: 1901-05, 8.
Bolton 7522.

Appendix: Ephemerides and Nautical Almanacs

0351. *La connoissance des temps; ou, Calendrier et éphémérides du lever et coucher du soleil, de la lune et des autres planètes.* Paris, 1679+. Founded and edited until 1684 by L. Dalencé (not by J. Picard, as is sometimes said). Edited from 1684 to 1701 by J. Le Fèvre. In 1702 it came under the control of the Académie Royale des Sciences and the title became *Connaissance des temps au méridien de Paris* (with later variations). Subsequent editors were appointed by the Academy. They were: 1702-29, J. Lieutaud; 1730-33, L. Godin; 1735-59, G.D. Maraldi; 1760-75, J.J. de Lalande; 1776-87, E.S. Jeaurat; 1788-94, P.F.A. Méchain. In 1795 control passed to the newly formed Bureau des Longitudes. From 1760, under the editorship of Lalande, its contents were expanded and in 1766 it began to include astronomical news and memoirs. Size: 1800-63, 1; 1864-73, 1. Continues to the present.
See Index. Scudder 1368. Bolton 1337. Documentation: DSB--articles on Dalencé, Le Fèvre, and Jeaurat, with some further information in the articles on other editors.

0352. *Ephémérides des mouvemens célestes.* Paris, 1716+. Edited 1716-34 (for the period 1715-44) by P. Desplaces, 1744-63 (for the period 1745-75) by N.L. de Lacaille, and 1774-92 (for the period 1775-1800) by J.J. de Lalande.
Scudder 1381. Bolton 1604.

0353. *Ephemerides astronomicae ad meridianum Vindobonensem.* Vienna, 1757+. Edited by M. Hell until his death in 1792, thereafter until its cessation in 1805 by F. Triesnecker and J.T. Bürg. Included numerous papers by Hell, his collaborators and other astromomers, as well as reprints of articles from other periodicals.
See Index. Scudder 3583. Bolton 1598. Documentation: Sarton, 1944; DSB--articles on Hell and Bürg.

0354. *The nautical almanac and astronomical ephemeris.* London, 1767+. Published by the Board of Longitude (under the superintendence of N. Maskelyne until his death in 1811, and then of T. Young). From 1832 published by the Admiralty. Continues to the present.
Scudder 393. Bolton 3287. Documentation: DSB--articles on Maskelyne and C. Mason.

0355. *Effemeridi astronomiche, calcolate al meridiano di Milano.* Also entitled *Ephemerides astronomicae....* Milano, 1774+. Edited by G.A. Cesaris (head of the Milan Observatory). Cont'd from 1806 as *Effemeridi astronomiche di Milano*, published by the Reale Osservatorio di Brera. Included astronomical memoirs. Size: 1800-63, 1.
See Index. Scudder 1889, 1904. Bolton 1531.

0356. (a) *Astronomisches Jahrbuch oder Ephemeriden, nebst einer Sammlung der neuesten in die astronomischen Wissenschaften einschlagenden Beobachtungen, Nachrichten, Bemerkungen und Abhandlungen.* Berlin, 1776 (1774)+. Published by the Berlin Academy and edited by J.E. Bode until his death in 1826. In this period it included astronomical papers and

observations. Size: 1800-29, 2. Cont'd as (b).
(b) *Berliner astronomisches Jahrbuch*. 1830 (1828)+. Published by the Berlin Academy and edited by J.F. Enke until his death in 1865; published thereafter by the Berlin Observatory. After 1829 (1825) it no longer included papers.
See Index. Scudder 2370b. Bolton 716. Documentation: Herrmann, 1972; DSB--articles on Bode, Encke, and Powalky.

0357. *Almanaque náutico y efemérides astronómicas*. Madrid, 1792 (1791)+. Published by the Observatorio Real de Marina de San Fernando. Cont'd past 1900.
Scudder 1730. Bolton 166.

0358. Bureau des Longitudes. Paris. (Founded 1795)
Annuaire. 1799 (1796)+. "Augmenté de notices scientifiques." Cont'd past 1900.
Scudder 1354. Bolton 477.

0359. *Ephemerides astronomicas calculadas para o meridiano do Observatorio real da Universidade de Coimbra*. Coimbra, 1804 (1803)+. Cont'd past 1900.
Scudder 1715. Bolton 1599.

0360. *Nautisches Jahrbuch, oder vollständige Ephemeriden und Tafeln*.... Berlin, 1852 (1850)+. Published by the Prussian government and edited until 1877 by C. Bremiker.
Scudder 2373. Bolton 3294.

0361. *The American ephemeris and nautical almanac*. Washington, 1855 (1852)+. Published by the U.S. Navy Department. The first superintendent was C.H. Davis, who was succeeded in 1877 by S. Newcomb. Continues to the present.
Scudder 4337. Bolton 202.

PHYSICS

Including some periodicals for telegraphy and electrical engineering.

Some of the periodicals listed under Chemistry also contained physics, notably 0393b, 0397d, and 0422.

0362. (a) *Journal der Physik*. Halle/Leipzig, 1790+. Edited by F.A.C. Gren. Cont'd from 1795 as *Neues Journal der Physik*. Cont'd as (b).
(b) *Annalen der Physik*. Ib., 1799+. Edited by L.W. Gilbert. Size: 1799-1824, 9. Cont'd from 1819 as *Annalen der Physik und der physikalischen Chemie*. Cont'd as (c).
(c) *Annalen der Physik und Chemie*. Leipzig, 1824+. Edited by J.C. Poggendorff. Size: 1824-63, 11; 1864-73, 11. Cont'd as (d).
(d) *Annalen der Physik und Chemie*. Neue Folge. Unter Mitwirkung der Physikalischen Gesellschaft in Berlin. Leipzig, 1877+. Edited by G. Wiedemann. Size: 1901-05, 12. (The words "und Chemie" were dropped from the title in 1900.) See also 0373.
See Index. Scudder 2718, 2704. Bolton 2410. Documentation: Schimank, 1963. DSB--articles on Gren and Poggendorff.

0363. *Zeitschrift für Physik und Mathematik*. Wien, 1826+. Edited by A. Baumgartner and A. von Ettingshausen. Size: 1826-32, 2. Cont'd from 1832 as *Zeitschrift für Physik und verwandte Wissenschaften*. Cont'd from 1840 as *Wiener Zeitschrift für Physik, Chemie und Mineralogie*. Size: 1832-41, 1. Ceased 1842.
Scudder 3647. Bolton 4873.

0364. (a) *Repertorium der Experimental-Physik. Enthaltend eine vollständige Zusammenstellung der neueren Fortschritte dieser Wissenschaft*. Leipzig, 1832. Edited by G.T. Fechner. Only three parts appeared; in 1836 it was revived as (b).
(b) *Neues Repertorium der Experimental-Physik*. Berlin, 1836. Edited by H.W. Dove and L. Moser. Cont'd after only one volume as (c).
(c) *Repertorium der Physik. Enthaltend eine vollständige Zusammenstellung der neueren Fortschritte dieser Wissenschaft*. Ib., 1837+. Edited by H.W. Dove and L. Moser. Ceased 1849.
See Index. Scudder 2999, 2360, 2381. Bolton 3920.

0365. (a) *Il cimento. Giornale di fisica, chimica e storia naturale*. Pisa, 1844+. Edited by C. Matteucci et al. Ceased 1846 and then revived as (b).
(b) *Il nuovo cimento. Giornale di fisica, di chimica e delle loro applicazioni*. (Sub-title varies) Ib., 1855+. Edited by C. Matteucci and R. Piria. Size: 1855-61, 6; 1901-05, 3 (all physics).
See Index. Scudder 2003, 2010. Bolton 1944.

0366. *Die Fortschritte der Physik*. Berlin, 1845 (1847)+. Published by the Physikalische Gesellschaft zu Berlin (founded 1843; see also 0381). An annual survey of the literature. Cont'd after 1918 as *Physikalische Berichte*.
Scudder 2365. Bolton 1765.

0367. *Annales télégraphiques*. Paris, 1855+. Size: 1855-63, 2. Ceased 1899.
Scudder 1311. Bolton 342 & p. 626.

0368. *Repertorium für physikalische Technik, für mathematische und astronomische Instrumentenkunde*. München, 1865+. Edited by P. Carl. Cont'd from 1868 as *Repertorium für Experimental-Physik*. Size: 1866-73, 3. Cont'd from 1883 as *Repertorium der Physik*. Edited by F. Exner. Ceased 1891.
Scudder 3125. Bolton 3932 & p. 908.

0369. *Telegraphic journal and electrical review*. London, 1872+. Edited by W. Higgs. Cont'd from 1892 as *The electrical review*. Size: 1901-05, 2.
Scudder 467. Bolton 4460 & p. 958.

0370. *Journal de physique théorique et appliquée*. Paris, 1872+. Edited by J.C. d'Almeida. Size: 1872-73, 11; 1901-05, 6.
See Index. Scudder 1422. Bolton 2388 and p. 795.

0371. Société Française de Physique. Paris. (Founded 1873)
Séances. 1873 (1874)+. Size: 1901-05, 3.
See Index. Scudder 1578.

0372. Physical Society of London. (Founded 1874) See also 0389.
Proceedings. 1874/75 (1876)+. Size: 1901-05, 1.
See Index. Scudder 413.

0373. (a) *Annalen der Physik und Chemie* [0362d] *Beiblätter*. Leipzig, 1877+. In 1920 it was absorbed in the *Physikalische Berichte* (0366).
(b) ——— [Supplement] *Literatur-Übersicht*. 1884-1903. Monthly lists of "current contents" of periodicals in the field.

0374. *The electrician. A weekly journal of theoretical and applied electricity and chemical physics*. (Sub-title varies) London, 1878+. Size: 1901-05, 2.
Bolton 1549.

0375. *La lumière électricique. Revue de l'electricité*. Paris, 1879+. Edited by T. Du Moncel. Cont'd from 1894 as *L'éclairage électrique*. Size: 1901-05, 7.
Bolton 2720.

0376. *Central-Zeitung für Optik und Mechanik*. Leipzig, 1880+. Edited by O. Schneider. Size: 1901-05, 4.
See Index. Bolton 1241.

0377. *Elektrotechnische Zeitschrift*. Organ des Elektrotechnischen Vereins. Berlin, 1880+. Size: 1901-05, 11.
Bolton 1556.

0378. *Elektrichestvo*. [*Electricity*] St. Petersburg, 1880+. Size: 1901-05, 1.
Bolton 6089.

0379. *L'électricien. Revue générale de l'électricité*. Paris, 1881+. Size: 1901-05, 11.
Bolton 1551 & p. 722/3.

0380. *Zeitschrift für Instrumentenkunde*. Berlin, 1881+. Edited by E. Abbe (*later* in association with the Physikalisch-technische Reichanstalt). Size: 1901-05, 5 (physics 59%, astronomy 24%, other 17%).
Bolton 4857 & p. 995.

0381. Physikalische Gesellschaft zu Berlin. (Founded 1843) See also 0366. From 1899: Deutsche Physikalische Gesellschaft.
Verhandlungen. 1882+. Size: 1901-05, 6.

0382. *Il giorno*. Milano, 1882+. Cont'd from 1886 as *L'elettricità*. Size: 1901-05, 1.
Bolton 6408.

0383. Rijksuniversiteit te Leiden. Natuurkundig Laboratorium (founded 1856); from 1882? Cryogeen Laboratorium.
Communications from the Physical Laboratory at the University of Leiden. 1885+. Size: 1901-05, 1. (In 1882 H. Kamerlingh Onnes was appointed professor of experimental physics at Leiden and soon became well known for his cryogenic investigations.)

0384. *Vestnik opytnoi fiziki i elementarnoi matematik*. [*Messenger of experimental physics and elementary mathematics*] Kiev, 1886+. Size: 1901-05, 3. Ceased 1908?

0385. *Zeitschrift für den physikalischen und chemischen Unterricht*. Berlin, 1887+. Edited by F. Poske. Size: 1901-05, 9 (physics 63%, mechanics 16%, chemistry 13%, other 8%).
Bolton 8453.

0386. *Revue pratique de l'électricité*. Paris, 1892+. Edited by A. Mandeix. Size: 1901-05, 2.
Bolton 7773.

0387. *L'industrie électrique. Revue de la science électrique et de ses applications industrielles*. Paris, 1892+. Edited by E. Hospitalier. Size: 1901-05, 3.
Bolton 6579.

0388. *The physical review. A journal of experimental and theoretical physics*. Ithaca, N.Y., 1893 (1894)+. Edited by E.L. Nichols and E. Merritt. "Published for Cornell University." Later: "Conducted with the cooperation of the American Physical Society." Size: 1901-05, 7.
See Index. Bolton 7502.

0389. (a) Physical Society of London. See also 0372.
Abstracts of physical papers from foreign sources. 1895-97. Succeeded by (b).
(b) *Science abstracts: Physics and electrical engineering*. "Issued under the direction of the Institution of Electrical Engineers [and] the Physical Society of London." 1898+. (Divided in 1903 into Section A for physics and Section B for electrical engineering)

0390. *Physikalische Zeitschrift*. Leipzig, 1899+. Size: 1901-05, 12.

CHEMISTRY

Including many pharmacy periodicals before 1850, and in the late nineteenth century some periodicals for applied chemistry which contained substantial numbers of papers relevant to pure chemistry.

In the 1790s and the early nineteenth century a good deal of chemistry was included in the *Journal der Physik*, later the *Annalen der Physik (und Chemie)* (0362).

0391. [The periodicals of Lorenz Crell]
(a) *Chemisches Journal für die Freunde der Naturlehre, Arzneygelahrtheit, Haushaltungskunst und Manufacturen*. Lemgo, 1778-81. 6 parts. Continued as (b).
(b) *Die neuesten Entdeckungen in der Chemie*. Leipzig, 1781-84. 12 vols. Continued as (c). (However another volume appeared in 1786)
(c) *Chemische Annalen für die Freunde der Naturlehre, Arzneygelahrtheit, Haushaltungskunst und Manufacturen*. Helmstädt/Leipzig, 1784-1803. 40 vols. Size: 1800-03, 4.
See Index (under Crell). Scudder 2801, 2982, 3054. Bolton 1254, 1258. Documentation: Geus, 1971; DSB--article on Crell.
Crell also published two short-lived auxiliary journals, *Chemisches Archiv* and *Beyträge zu den Chemischen Annalen*.

0392. (a) *Almanach für Scheidekünstler und Apotheker*. Weimar, 1780+. Edited by J.F.A. Göttling. Continued as (b).
(b) *Taschenbuch für Scheidekünstler und Apotheker*. Ib., 1803+.

Edited by C.F. Bucholz. Cont'd as *Chemisches Taschenbuch für Ärzte, Chemiker und Pharmaceuten*. Then cont'd as (c).
(c) *Trommsdorff's Taschenbuch für Ärzte, Chemiker und Pharmaceuten*. Jena, 1820+. Edited by J.B. Trommsdorff. Ceased 1829.
See Index. Scudder 3296. Bolton 153, 4424.

0393. (a) *Annales de chimie*. Paris, 1789+. Edited by Guyton de Morveau, Lavoisier, Monge, Berthollet, Fourcroy, et al. Size: 1789-1815, 8. Cont'd as (b).
(b) *Annales de chimie et de physique*. Ib., 1816+. Through the 19th century it was conducted by an editorial board comprising many of the leading French chemists of the time with some physicists. Size: 1816-40, 11; 1841-63, 10; 1864-73, 11; 1901-05, 7 (chemistry 70%, physics 30%).
See Index. Scudder 1292. Bolton 346, 418, 761. Documentation: Court, 1972; Delépine, 1962.

0394. *Annali di chimica (*later *e storia naturale)*. Pavia, 1790+. Edited by L. Brugnatelli. Ceased 1802.
Scudder 1990. Bolton 403.

0395. *Journal der Pharmacie für Aerzte und Apotheker (*later *und Chemisten)*. Leipzig, 1794+. Edited by J.B. Trommsdorff. Size: 1794-1816, 3. Cont'd from 1817 as *Neues Journal der Pharmacie* .... Size: 1817-33, 5. In 1835 it was absorbed in the *Annalen der Pharmacie* (0409).
See Index (under Trommsdorff). Scudder 2979. Bolton 2409. Documentation: DSB--article on Trommsdorff.

0396. *Berlinisches Jahrbuch für die Pharmacie und die damit verbundenen Wissenschaften*. Berlin, 1795+. Cont'd from 1803 as *Neues Berlinisches Jahrbuch*.... Edited from 1803 to 1810 by A.F. Gehlen and V. Rose. Later editors included J.W. Döbereiner and C.W.G. Kastner. Ceased 1840.
Scudder 2293. Bolton 902.

0397. (a) *Allgemeines Journal der Chemie*. Leipzig, 1798+. Edited by A.N. Scherer. Size: 1798-1802, 5. Cont'd as (b).
(b) *Neues allgemeines Journal der Chemie*. Berlin, 1803+. Edited by A.F. Gehlen. Size: 1803-06, 7. Cont'd as (c).
(c) *Journal für die Chemie und Physik (*later *und Mineralogie)*. Berlin, 1806+. Edited by A.F. Gehlen. Size: 1806-10, 10. Cont'd as (d).
(d) *Journal für Chemie und Physik*. Nürnberg, 1811+. Edited by J.S.C. Schweigger. Size: 1811-33, 11. In 1834 it merged with the *Journal für technische und ökonomische Chemie* (0407a) to form the *Journal für praktische Chemie* (0407b).
See Index. Scudder 2882, 2338, 3167. Bolton 128 & p. 613. Documentation: DSB--articles on Gehlen and Schweigger.

0398. *Afhandlingar i fysik, kemi och mineralogi*. Stockholm, 1806+. Edited by W. Hisinger and J. Berzelius. Size: 1806-18, 3. Most of the papers were by Berzelius. Ceased 1818.
Scudder 676. Bolton 39.

0399. (a) *Bulletin de pharmacie (*later *et des sciences accessoires)*. Paris, 1809+. Edited by A.A. Parmentier, C.L. Cadet, et al. Cont'd as (b).

(b) *Journal de pharmacie et des sciences accessoires* (Contenant le *Bulletin des travaux* de la Société de Pharmacie de Paris). Ib., 1815+. Size: 1815-41, 7. Cont'd as (c).
(c) *Journal de pharmacie et de chimie* (Contenant....) Ib., 1842+. Size: 1842-63, 11; 1864-73, 10; 1901-05, 4.
Scudder 1348. Bolton 1133 & p. 680.

0400. *Repertorium für die Pharmacie*. Nürnberg, 1815+. "Angefangen von A.F. Gehlen und fortgesetzt ... von J.A. Buchner." Size: 1819-51, 2. Cont'd from 1852 as *Neues Repertorium....* Ceased 1876.
Bolton 3933.

0401. (a) *Årsberättelse om framstegen i physik och chemi (*from 1841 *i kemi och mineralogi*, from 1847 *i kemi)*. Stockholm, 1821 (1822)+. Published by the K. Vetenskaps-Akademien (see 074) and edited by J. Berzelius. Ceased 1849 (1851). The last three volumes (1847-49) were edited by L.F. Svanberg.
Scudder 693h,j,k. Bolton 654.
(b) ——— [Trans.] *Jahres-Bericht über die Fortschritte der physischen Wissenschaften (*later *der Chemie und Mineralogie*, later *der Chemie)*. Tübingen, 1822+. Trans. from the Swedish by C.G. Gmelin, later by F. Wöhler. Ceased 1850 (1851). For its successor see 0416.
See Index (under Berzelius). Scudder 3278. Bolton 2295.
(c) ——— [Trans.] *Rapport annuel sur les progrès des sciences physiques et chimiques (*later *de la chimie)*. Paris, 1841+. "Traduit du suédois sur les yeux de l'auteur." Trans. by P. Plantamour. Ceased 1848.
Scudder 1278c. Bolton 3861.

0402. *Archiv für die Pharmazie*. Schmalkalden, 1822+. Published by the Apotheker-Verein im nördlichen Teutschland and edited by R. Brandes. In 1832 it merged with the *Magazin für Pharmacie und Experimental-Kritik* (0403c) to form the *Annalen der Pharmacie* (0409). However from 1835 it continued independently under the editorship of Brandes as *Archiv der Pharmazie*. Size: 1826-63, 3; 1864-73, 2; 1901-05, 7.
Bolton 3572 & p. 885; cf. 3577.

0403. (a) *Magazin für die neuesten Erfahrungen, Entdeckungen und Berichtigungen im Gebiete der Pharmazie*. Karlsruhe, 1823+. Edited by G.F. Hänle. Cont'd as (b).
(b) *Magazin für Pharmacie und die dahin einschlagenden Wissenschaften*. Ib., 1824+. Edited by P.L. Geiger. Size: 1823-31, 5. Cont'd as (c).
(c) *Magazin für Pharmacie und Experimental-Kritik*. Heidelberg, 1831+. Edited by P.L. Geiger and J. Liebig. In 1832 it merged with the *Archiv für die Pharmazie* (0402) to form the *Annalen der Pharmacie* (0409).
Bolton 3342b.

0404. (a) *Giornale di farmacia, chimica e scienze accessorie*. Milano, 1824+. Edited by A. Cattaneo. Size: 1824-33, 2. Cont'd as (b).
(b) *Biblioteca di farmacia, chimica....* Ib., 1834+. Edited by Cattaneo. Size: 1834-45, 1. Cont'd as (c).

(c) *Annali di chimica applicata alla medicina....* [Sub-title varies] Ib., 1845+. Edited by G. Polli. Size: 1845-63, 2; 1864-73, 5. Ceased 1900.
Scudder 1895, 1883. Bolton 1970 & p. 755.

0405. *Journal de chimie médicale.* Paris, 1825+. Published by the Société de Chimie Médicale. Size: 1825-44, 2; 1845-63, 2; 1864-73, 1. In 1877 it was incorporated in the *Répertoire de pharmacie* (not included in the present catalogue).
Scudder 1568. Bolton 2366.

0406. *American journal of pharmacy.* Philadelphia, 1825+. Size: 1836-60, 1. Cont'd past 1900.
Bolton 2530.

0407. (a) *Journal für technische und ökonomische Chemie.* Leipzig, 1828+. Edited by O.L. Erdmann. Size: 1828-32, 11. In 1834 it united with the *Journal für Chemie und Physik* (0397d) to form (b).
(b) *Journal für praktische chemie.* Ib., 1834+. Edited by O.L. Erdmann until his death in 1869, then by H. Kolbe. Size: 1834-63, 12; 1864-69, 12; 1870-73, 11; 1901-05, 6.
See Index. Scudder 2950, 2948. Bolton 2487, 128d,e & p. 613.

0408. *Pharmaceutisches Central-Blatt.* Leipzig, 1830+. Size: 1830-49, 1. Cont'd from 1850 as *Chemisch-pharmaceutisches Central-Blatt.* Size: 1850-55. 1. Cont'd from 1856 as *Chemisches Central-Blatt.* Size: 1856-63, 2; 1864-73, 4. By the late 19th century it had become entirely an abstracting periodical. In 1897 it was taken over by the Deutsche Chemische Gesellschaft.
See Index. Scudder 2988. Bolton 3584. Documentation: Pflücke, 1954; Wilstätter, 1929.

0409. (a) *Annalen der Pharmacie.* Lemgo/Heidelberg, 1832+. Formed by union of 0402 and 0403c. Edited initially by R. Brandes, P.L. Geiger, and J. Liebig, then from 1838 by J. Liebig and F. Wöhler. Cont'd from 1840 as *Annalen der Chemie und Pharmacie.* Size: 1832-63, 11; 1864-73, 12. Cont'd after Liebig's death in 1873 as (b).
(b) *Justus Liebig's Annalen der Chemie und Pharmacie.* Leipzig/ Heidelberg, 1873+. The words "und Pharmacie" were dropped from the title in 1874. Size: 1901-05, 9.
See Index. Scudder 3053. Bolton 321. Documentation: Van Klooster, 1957.

0410. (a) *Revue scientifique et industrielle.* Paris, 1840+. Edited by G.A. Quesneville. Succeeded by (b).
(b) *Moniteur scientifique du chimiste et du manufacturier.* Ib., 1857+. Cont'd from 1871 as *Moniteur scientifique de Quesneville.* Size: 1857-73, 3; 1901-05, 3.
Scudder 1536, 1456. Bolton 4055.

0411. *The chemist; or, Reporter of chemical discoveries and improvements (in analytical, manufacturing and agricultural chemistry).* [Sub-title varies] London, 1840+. Edited by C. Watt and J. Watt. Size: 1840-58, 3. Ceased 1858.
Scudder 240. Bolton 1263.

0412. Chemical Society of London. (Founded 1841) See also 0435.
(a) *Memoirs (*later *and Proceedings)*. 1841/43 (1843)+. Size: 1841-48, 4. Cont'd as (b).
(b) *Quarterly journal*. 1849+. Size: 1849-63, 5. Cont'd as (c).
(c) *Journal*. [Issued monthly] 1862+. Size: 1864-73, 11; 1901-05, 11.
See Index. Scudder 236.

0413. *Pharmaceutical journal (and transactions)*. London, 1841+. Edited by J. Bell. Size: 1841-63, 5; 1864-73, 5; 1901-05, 3.
Bolton 3568 & p. 884.

0414. *The chemical gazette; or, Journal of practical chemistry in all its applications to pharmacy, arts and manufactures*. London, 1842/43 (1843)+. Edited by W. Francis. Size: 1842-54, 1. In 1859 it was absorbed in *The chemical news* (0419).
Scudder 232. Bolton 1247.

0415. Utrechtsche Hoogeschool [Utrecht University]. See also 0875.
Scheikundige onderzoekingen, gedann in het laboratorium. [Title varies] Rotterdam, 1842+. Edited by G.J. Mulder. Size: 1842-51, 3; 1854-56, 1; 1857-64, 1. Ceased 1865.
Scudder 905, 924, 907. Bolton 4215. Documentation: DSB--article on Mulder.

0416. (a) *Jahresbericht über die Fortschritte der reinen, pharmaceutischen und technischen Chemie, Physik, Mineralogie und Geologie*. Giessen, 1847 (1849)+. Edited by J. Liebig and H. Kopp. Founded as a successor to Berzelius' *Jahresbericht* (0401b) which Berzelius had relinquished in 1846, shortly before his death. Cont'd as (b).
(b) *Jahresbericht über die Fortschritte der Chemie und verwandter Theile anderer Wissenschaften*. Ib., 1857 (1858)+. Edited by H. Kopp et al. Ceased 1910.
Scudder 2640, 2638. Bolton 2296.
(c) [Trans. of (a)] *Annual reports of the progress of chemistry and the allied sciences*. London, 1847-53 (1849-55). Trans. by A.W. Hoffmann and W. De La Rue.
Scudder 197. Bolton 483.

0417. Société Chimique de Paris (*from 1907* de France).
(a) *Bulletin*. 1858/60+. Size: 1858-63, 4; 1864-73, 10; 1901-05, 12.
See Index. Scudder 1569. Bolton 3909a.
(b) *Répertoire de chimie pure*. 1858/59+. Size: 1859-62, 1. Absorbed in (a) in 1863.
Scudder 1507. Bolton 3909a.
(c) *Répertoire de chimie appliquée*. 1858/59+. Size: 1859-63, 7. Absorbed in (a) in 1864.
Scudder 1506. Bolton 3909b.

0418. (a) *Kritische Zeitschrift für Chemie, Physik und Mathematik (*from 1859 *für Chemie und die verwandten Wissenschaften)*. Erlangen, 1858+. Edited by A. Kekulé et al., then from 1859 by E. Erlenmeyer. Cont'd as (b).
(b) *Zeitschrift für Chemie*. Leipzig, 1860+. Edited from 1865 by F. Beilstein, R. Fittig, and H. Hübner. Size: 1861-63, 7; 1864-71, 11. Ceased 1871.
Scudder 2568. Bolton 2583.

0419. *Chemical news (*later *and journal of physical science).* London, 1859/60+. Edited by W. Crookes. Size: 1860-63, 11; 1864-73, 10; 1901-05, 6. Ceased 1932.
Scudder 233, 4189. Bolton 1247a.

0420. *Zeitschrift für analytische Chemie.* Wiesbaden, 1862+. Edited by C.R. Frescenius. Size: 1862-73, 4; 1901-05, 7.
Scudder 3322. Bolton 4837 & p. 992.

0421. Deutsche Chemische Gesellschaft. Berlin. (Founded 1867)
Berichte. 1868+. Size: 1868-73, 12; 1901-05, 17.
See Index. Scudder 2311. Documentation: Hückel, 1967.

0422. Russkoe Khimicheskoe (*from 1873* Russkoe Fiziko-Khimicheskoe) Obshchestvo. (Société Physico-Chimique Russe) St. Petersburg.
Zhurnal. 1869+. Size: 1901-05, 12 (chemistry 81%, physics 17%, other 2%).
Scudder 3742. Bolton 8475.

0423. *Jahresberichte über die Fortschritte der Thierchemie, oder der physiologischen und pathologischen Chemie.* [Sub-title varies] Wien, 1871+. Edited by R. Maly. An abstracting and review journal. Ceased 1919.
Scudder 3593. Bolton 2297, 6691.
"A useful survey of the beginnings of biochemistry which is conveniently provided with cumulative decennial indexes."

0424. *La gazzetta chimica italiana.* Palermo, 1871+. Edited by E. Paternò. Size: 1871-73, 10; 1901-05, 6. Later taken over by the Società Chimica Italiana.
Scudder 1977. Bolton 1846 & p. 750.

0425. *Listy chemické* [*Chemical journal*]. Praha, 1875+. Published by the Spolek Chemiků Českých [Society of Czech Chemists]. Size: 1901-05, 2.
Bolton 2698 & p. 819.

0426. *The analyst.* London, 1876+. Published by the Society of Public Analysts. Size: 1901-05, 3.
Scudder 454. Bolton 289.

0427. *Zeitschrift für physiologische Chemie.* Strassburg, 1877+. Edited by F. Hoppe-Seyler. Cont'd after his death in 1895 as *Hoppe-Seyler's Zeitschrift für physiologische Chemie.* Size: 1901-05, 12 (chemistry 50%, physiology 43%, other 7%).
Bolton 4875 & p. 997.

0428. *Allgemeine Chemiker-Zeitung.* Cöthen, 1877+. Edited by G. Krause. Cont'd from 1879 as *Chemiker-Zeitung.* Size: 1901-05, 12.
Bolton 88 & p. 609. Documentation: Maas, 1976.

0429. *American chemical journal.* Baltimore, 1879+. Edited by I. Remsen. Size: 1901-05, 6. In 1914 it was incorporated in 0430.
See Index. Bolton 192 & p. 617.

0430. American Chemical Society. (Founded 1876)
Journal. Washington, later Easton, Pa., 1879+. Size: 1901-05, 11.
See Index.

0431. *Monatshefte für Chemie.* Wien, 1880+. Published by the Akademie

der Wissenschaften in Wien. Size: 1901-05, 8.
Bolton 3108 & p. 847.

0432. Tokyo Chemical Society. (Founded 1878)
Journal (Tokyo kwagaku kwai shi). 1880+. Size: 1901-05, 2.
Bolton 8215.

0433. Society of Chemical Industry. London.
Journal. 1882+. Size: 1901-05, 9.

0434. *Recueil des travaux chimiques des Pays-Bas.* Leiden, 1882+. Size: 1901-05, 4. Taken over in 1920 by the Société Chimique Néerlandaise.
Bolton 3887.

0435. Chemical Society of London. See also 0412.
Abstracts of the proceedings. 1885+. Cont'd from 1890 as *Proceedings.* Size: 1901-05, 7.

0436. *Vierteljahrsschrift über die Fortschritte auf dem Gebiete der Chemie der Nahrungs- und Genussmittel.* Berlin, 1886+. Cont'd from 1898 as *Zeitschrift für Untersuchung der Nahrungs- und Genussmittel.* Size: 1901-05, 10 (chemistry 78%, physiology 22%).
Bolton 8288.

0437. *Zeitschrift für physikalische Chemie.* Leipzig, 1887+. Edited by W. Ostwald and J.H. van't Hoff. Size: 1901-05, 11.
Bolton 8440.

0438. *Kemiska notiser.* Stockholm, 1887+. Published by the Kemistssamfundet [Chemists' Association] i Stockholm. Cont'd from 1889 as *Svensk kemisk tidskrift.* Size: 1901-05, 3.
Bolton 6813.

0439. *Zeitschrift für angewandte Chemie.* Berlin, 1888+. Published by the Deutsche Gesellschaft für Angewandte Chemie. Size: 1901-05, 11.
Bolton 8454a.

0440. *Zeitschrift für anorganische Chemie.* Hamburg, 1892+. Edited by G. Krüss. Size: 1901-05, 11.
Bolton 8405.

0441. *Jahrbuch der Chemie: Bericht über die neuesten und wichtigsten Fortschritte der reinen und angewandten Chemie.* Frankfurt a.M., 1892+. Edited by R. Meyer. Ceased 1918.
Bolton 6663.

0442. International Congress of Pure and Applied Chemistry. [Also in French, German, Italian, and Spanish] (The first meeting was held at Brussels in 1894)
Proceedings. 1894+.

0443. *Jahrbuch der Elektrochemie (*later *und angewandten physikalischen Chemie).* Halle, 1894 (1895)+. Edited by W. Nernst and W. Borchers. Ceased 1909.
Bolton 6664.

0444. *Zeitschrift für Elektrochemie.* Leipzig, 1894+. Published by the Deutsche Elektrochemische Gesellschaft (from 1902 the Deutsche Bunsen-Gesellschaft für Angewandte Physikalischen

Chemie). Edited by W. Borchers in collaboration with W. Ostwald. Size: 1901-05, 11 (chemistry 73%, physics 24%).
Bolton 8410.

0445. *Elektrochemische Zeitschrift*. Berlin, 1894+. Edited by Klobükow. Size: 1901-05, 4 (chemistry 65%, physics 35%). Ceased 1922.
Bolton 6091.

0446. *Review of American chemical research*. Cambridge, Mass. (later Easton, Pa.), 1895+. Edited by A.A. Noyes. In 1907 it became *Chemical abstracts*.

0447. *Centralblatt für Nahrungs- und Genussmittel-Chemie*. Weimar, 1895+. Cont'd from 1897 as *Zeitschrift für öffentliche Chemie*. Size: 1901-05, 5 (chemistry 83%, physiology 17%).
Bolton 8434.

0448. *L'électrochimie: Revue des sciences et de l'industrie*. Paris, 1895+. Edited by A. Minet. Size: 1901-05, 3 (physics 76%, chemistry 24%).
Bolton 6085.

0449. *Magyar chemiai folyóirat [Hungarian chemical journal]*. Budapest, 1895+. Size: 1901-05, 3.

0450. *The journal of physical chemistry*. Ithaca, N.Y., 1896+. Edited by W.D. Bancroft. "Published at Cornell University." Size: 1901-05, 5.
Bolton 6787.

0451. *Annales de chimie analytique*. Paris, 1896+. Edited by C. Crinon. Size: 1901-05, 7.
Bolton 8483.

0452. *Sammlung chemischer und chemisch-technischer Vorträge*. Stuttgart, 1896+. Edited by Ahrens. Size: 1901-05, 2.

0453. *Actualités chimiques: Revue des progrès de la chimie pure et appliquée*. Paris, 1896+. Edited by C. Friedel. Cont'd from 1899 as *Revue générale de chimie pure et appliquée*.

GEOLOGY

Including palaeontology and mineralogy, and some early periodicals for mining.

Titles of a large number of periodicals of geological surveys were found but most of them were not cited in the Royal Society's *Catalogue of Scientific Papers* or the *International Catalogue of Scientific Literature* and so were not included. (Information about the publications of geological surveys can be found in Scudder and standard bibliographical sources under the names of countries and states.)

0454. *Bergmännisches Journal*. Freiberg, 1788+. Edited by A.W. Köhler and C.A.S. Hoffmann. Cont'd from 1795 as *Neues bergmännisches Journal*. Size: 1795-1815, 1. Ceased 1815.
See Index. Scudder 2617, 2625. Bolton 864.

0455. (a) *Journal des mines*. Publié par l'Agence (*later* le Conseil) des Mines de la République. Paris, 1794+. Size: 1794-1815, 5. Cont'd as (b).
(b) *Annales des mines....* Rédigées par le Conseil Général des Mines. Paris, 1816+. Size: 1817-63, 4; 1864-73, 2; 1901-05, 1.
See Index. Scudder 1418, 1302. Bolton 2434.

0456. (a) *Taschenbuch für die gesammte Mineralogie*. Frankfurt a.M., 1807+. Edited by K.C. von Leonhard. Size: 1807-24, 1. From 1825 also known as *Zeitschrift für Mineralogie*. Cont'd as (b).
(b) *Jahrbuch für Mineralogie, Geognosie, Geologie und Petrefaktenkunde*. Heidelberg, 1830+. Cont'd from 1833 as *Neues Jahrbuch....* Edited by von Leonhard until his death in 1862. Size: 1833-63, 5; 1864-73, 6; 1901-05, 5.
See Index. Scudder 2608, 2793. Bolton 4416. Documentation: Bauer, 1907; DSB--article on von Leonhard.

0457. Geological Society of London. (Founded 1807) See also 0462.
Transactions. 1807/11 (1811)+. Size: 1811-56, 1. Ceased 1856.
Scudder 293.

..... Wernerian Natural History Society. Edinburgh.
Memoirs. 1808 (1811)+. Main entry 0597. Largely geological.

0458. Jernkontor [Iron Office]. Stockholm.
Jernkontorets annaler: En tidskrift för svenska bergshandteringen. 1817+. Edited by J. Åkerman. Size: 1845-63, 2. Cont'd past 1900.
Scudder 683. Bolton 2316 & p. 792.

0459. Royal Geological Society of Cornwall. (Founded 1814)
Transactions. London/Penzance, 1818+. Size: 1818-60, 1; 1864-73, 1; 1901-05, 1.
Scudder 521.

0460. (a) *Archiv für Bergbau und Hüttenwesen*. Breslau, 1818+. Edited by C.J.B. Karsten. Size: 1818-31, 2. Cont'd as (b).
(b) *Archiv für Mineralogie, Geognosie, Bergbau und Hüttenkunde*. Berlin, 1829+. Size: 1829-55, 2. Ceased 1855.
Scudder 2275, 2276. Bolton 584.

0461. *Gornuy zhurnal*. (*Journal des mines de Russie*) St. Petersburg, 1825+. Edited by D.I. Sokolov. Size: 1901-05, 1.
Scudder 3724. Bolton 1996.

0462. Geological Society of London. See also 0457.
(a) *Proceedings*. 1826/33 (1834)+. Size: 1826-45, 2. After 1845 it was included in (b).
(b) *Quarterly journal*. 1845+. Size: 1845-63, 6; 1864-73, 7; 1901-05, 8.
Scudder 293.

0463. *Kalendar für den sächsischen Berg- und Hüttenmann*. Freyberg, 1827+. Published by the K. Bergacademie zu Freyberg. Cont'd from 1852 (or 1837?) as *Jahrbuch für den Berg- und Hüttenmann*. (Eine Übersicht der Fortschritte des gesammten Berg- und Hüttenwesens sammt seiner Literatur. Mit statistischen und personal Nachrichten, etc.) Size: 1854-61, 1. Cont'd from 1873

as *Jahrbuch für das Berg- und Hüttenwesen im Königreich Sachsen*. Size: 1901-05, 1. Ceased 1938.
Scudder 2616, 2623. Bolton 2545.

0464. Société Géologique de France. Paris. (Founded 1830) See also 0466.
Bulletin. 1830+. Size: 1830-42, 7; 1843-63, 8; 1864-73, 4; 1901-05, 9.
Scudder 1582.

0465. Imp. Mineralogicheskoe Obshchestvo. (Russisch-Kaiserliche Mineralogische Gesellschaft) St. Petersburg.
Trudui. 1830+. Cont'd from 1842 as *Schriften*, then *Verhandlungen*. Cont'd from 1866 as *Zapiski*. Size: 1842-58, 1; 1901-05, 2.
Scudder 3726, 3737, 3719.

0466. Société Géologique de France. See also 0464.
Mémoires. 1833+. Size: 1833-42, 2; 1844-56, 1; 1901-05, 1.
Scudder 1582.

0467. Geological Society of Dublin. *From 1864* Royal Geological Society of Ireland.
Journal. 1833+. Size: 1838-64, 2; 1864-73, 1. Ceased 1885 (1887).
Scudder 90.

0468. Geological and Polytechnic Society of the West Riding of Yorkshire. (Founded 1837) *From 1871* Yorkshire Geological and Polytechnic Society. *Later* Yorkshire Geological Society.
Proceedings. Leeds, 1839/42+. Size: 1839-63, 1; 1864-72, 1; 1901-05, 2.
Scudder 153.

0469. Manchester Geological Society. (Founded 1838)
Transactions. 1840/41 (1841)+. Size: 1901-05, 2.
Scudder 496.

0470. Association of American Geologists (*from 1843* and Naturalists). (Held annual meetings, 1840-47, then succeeded by the American Association for the Advancement of Science: see 0136)
Reports of the meetings (later *Abstracts of the proceedings)*. 1840-45. Size: 1840-45, 2.
Scudder 3939.

0471. *Berg- und hüttenmännische Zeitung. Mit besonderer Berücksichtigung der Mineralogie und Geologie*. Nordhausen, 1842+. Edited by C. Hartmann. Cont'd from 1859 as *Allgemeine berg- und* .... Size: 1842-63, 1; 1864-73, 2; 1901-05, 3.
Scudder 3143, 3185. Bolton 869.

0472. Geological Survey of Great Britain (*later* of England and Wales, *later* of the United Kingdom). See also 0535.
Memoirs of the Geological Survey ... and of the Museum of Economic (later *of Practical) Geology in London*. 1846+. Size: 1846-63, 1; 1901-05, 1.
Scudder 294.

0473. *Palaeontographica: Beiträge zur Naturgeschichte der Vorwelt*. Cassel, 1846 (1851)+. Edited by W. Dunker and H. von Meyer.

Size: 1863-71, 2; 1901-05, 1.
Scudder 2472. Bolton 3520.

0474. Deutsche Geologische Gesellschaft. Berlin. (Founded 1848)
Zeitschrift. 1848/49+. Size: 1849-62, 4; 1864-73, 6; 1901-05, 7.
Scudder 2313.

0475. K.-K. Geologische Reichanstalt. Wien. See also 0481.
Jahrbuch. 1850+. Size: 1850-63, 6; 1864-73, 2; 1901-05, 4.
Scudder 3588.

0476. *Revista mineraria. Periodico cientifico é industrial*. Madrid, 1850+. Size: 1850-63, 1. Cont'd past 1900.
Scudder 1757. Bolton 3981.

0477. K.-K. Montan-Lehranstalt zu Leoben. (The name of the issuing body varies widely and includes at different times the mining academies at Přibram, Schemnitz, and Vordernberg.)
Berg- und hüttenmännisches Jahrbuch. Wien, 1851+. Size: 1856-63, 1; 1864-73, 2; 1901-05, 1.
Scudder 3573.

0478. *Zeitschrift für das Berg-, Hütten- und Salinenwesen in dem preussischen Staate*. Berlin, 1853+. Size: 1858-63, 2; 1864-73, 2. Cont'd past 1900.
Scudder 2373b. Bolton 4896.

0479. *Oesterreichische Zeitschrift für Berg- und Hüttenwesen*. Wien, 1853+. Edited by O. von Hingenau. Size: 1864-73, 2. Ceased 1914.
Scudder 3628. Bolton 3443.

0480. Geological Survey of India. See also 0487.
Memoirs. Calcutta, 1856 (1859)+. Size: 1901-05, 1.
Scudder 3784. Bolton 6984.

0481. K.-K. Geologische Reichanstalt. Wien. See also 0475.
Verhandlungen. 1858+. (Until 1867 it was published with the *Jahrbuch*--0475) Size: 1858-63, 1; 1864-73, 3; 1901-05, 6.
Scudder 3588.

0482. (a) *The geologist. A popular magazine*. London, 1858+. Edited by S.J. Mackie. Size: 1858-64, 2. Succeeded by (b).
(b) *The geological magazine*. Ib., 1864+. Edited by T.R. Jones and H. Woodward. Size: 1864-73, 11; 1901-05, 11.
Scudder 295, 290. Bolton 1896 & p. 752.

0483. Liverpool Geological Society. (Founded 1859)
Abstract of the Proceedings. 1859+. Cont'd from 1874 as *Proceedings*. Size: 1863-73, 1; 1901-05, 2.
Scudder 163.

0484. Geological Society of Glasgow.
Transactions. 1860+. Size: 1864-73, 2; 1901-05, 1.
Scudder 134.

0485. *Revue de géologie*. Paris, 1860+. Edited by A. Delesse and A. Lapparent. A review periodical. In 1879 it was merged in the *Annales des mines*.
Scudder 1524. Bolton 4074.

0486. The Geologists' Association. London.

Proceedings. 1859/65? (1865)+. Size: 1865-74, 1; 1901-05, 3. See Index. Scudder 297.

0487. Geological Survey of India. See also 0480.
Records. Calcutta, 1868 (1870)+. Size: 1870-73, 2; 1901-05, 1.
Scudder 3784. Bolton 7647.

0488. *Annales des sciences géologiques*. Paris, 1869+. Edited by E. Hébert and A. Milne-Edwards. Size: 1869-73: 1. In 1892 it was absorbed in *Annales des sciences naturelles* (0689 and 0758).
Scudder 1305. Bolton 375.

0489. Geological Survey of Indiana.
Annual report. 1869+. Size: 1901-05, 1.
Scudder 4092.

0490. Edinburgh Geological Society.
Transactions. 1866/69 (1870)+. Size: 1870-74, 3; 1901-05, 1.
Scudder 110.

0491. Reale Comitato Geologico d'Italia.
Bollettino. Firenze, later Roma, 1870+. Size: 1870-73, 4; 1901-05, 1.
Scudder 1832.

0492. Société Géologique du Nord. Lille.
Annales. 1870+. Size: 1901-05, 2.
Scudder 1166.

0493. *Földtani közlöny*. *(Geologische Mittheilungen)* Budapest, 1871 (1872)+. Published by the Magyarhoni Földtani Társulat (Ungarische Geologische Gesellschaft). Size: 1901-05, 2.
Scudder 3369. Bolton 1732.

0494. *Mineralogische Mittheilungen*. Wien, 1871+. Edited by G. Tschermak. Size: 1871-73, 6. Up to 1877 it was issued as a supplement to the *Jahrbuch* of the K.-K. Geologische Reichanstalt (0475). Cont'd from 1878 as *Mineralogische und petrographische Mittheilungen*. Cont'd from 1889 as *Tschermaks mineralogische und....* Size: 1901-05, 4.
Scudder 3604. Bolton 3014 & p. 840.

0495. Geologiska Föreningen [Geological Society] i Stockholm.
Förhandlingar. 1872+. Size: 1901-05, 3.
Scudder 679.

0496. Société Géologique de Belgique. (Cf. 0513)
Annales. Liége, 1874+. Size: 1901-05, 1.
Scudder 981.

0497. Schweizerische Paläontologische Gesellschaft.
Abhandlungen. Basel, etc., 1874+. Size: 1901-05, 1.
Scudder 2210.

0498. *The geological record: An account of works on geology, mineralogy and palaeontology published during the year*. [Sub-title varies] London, 1875+. Edited by W. Whitaker. Ceased 1889.
Scudder 292. Bolton 1894.

0499. Mineralogical Society of Great Britain and Ireland.
Mineralogical magazine and journal. Truro, later London, 1876+.

Size: 1901-05, 2.
Scudder 23. Bolton 3010.

0500. *Meddeleleser om Grønland*. Kjøbenhavn, 1876+. Published by the Commission for Ledelsen af de Geologiske og Geografiske Undersøgelser i Grønland. (Summaries in French) Size: 1901-05, 2.
Bolton 6969.

0501. (a) *Zeitschrift für Krystallographie und Mineralogie*. Leipzig, 1877+. Edited by P. Groth. Size: 1901-05, 7.
Bolton 4858 & p. 995.
(b) ——— [Supplement] *Repertorium der mineralogischen und krystallographischen Literatur*. Ib., 1876 (1877)+. Edited by P. Groth. It also included an index to (a). Ceased 1902.

0502. International Geological Congress. [Also in French, Italian, and German] (The first meeting was held in Paris in 1878)
Proceedings. 1878+. Size: 1901-05, 4.

0503. Société Minéralogique de France. (*From 1886* Société Française de Minéralogie)
Bulletin. Meulan, later Paris, 1878+. Size: 1901-05, 2.

0504. Geological Survey of the United States.
Various publications, the first of which began in 1880. Size (combined): 1901-05, 9.

0505. K. Preussische Geologische Landesanstalt und Bergakademie zu Berlin.
Jahrbuch. 1880+. Size: 1901-05, 3.

0506. *Beiträge zur Paläontologie (*from 1894 *und Geologie) Österreich-Ungarns und des Orients*. Wien, 1881+. Edited by E. von Mojsisovics and M. Neumayr. From 1894 sub-titled "Mittheilungen des Paläontologischen Instituts der Universität Wien." Size: 1901-05, 1.
Bolton 839.

0507. Rijks Geologisch-Mineralogisch Museum. Leiden.
Sammlungen. 1881+. Size: 1901-05, 2.
Bolton 7903.

0508. *Palaeontologische Abhandlungen*. Jena, 1882+. Edited by W. Dames and E. Kayser. Size: 1901-05, 1.
Bolton 7407.

0509. Società Geologica Italiana. Roma. (Founded 1881)
Bollettino. 1882+. Size: 1901-05, 4.

0510. Geologicheskago Komiteta. (Comité Géologique) St. Petersburg. See also 0511.
Izvestiya (Bulletin). 1882+. Size: 1901-05, 3.

0511. Geologicheskago Komiteta. See also 0510.
Trudui (Mémoires). 1883+. Size: 1901-05, 1.

0512. Geological (*from 1881* and Natural History) Survey of Canada. (Publication record dates from 1844)
Summary report. Montreal, 1885+. Size: 1901-05, 1.

0513. Société Belge de Géologie, de Paléontologie et d'Hydrologie.

Bruxelles. (Cf. 0496)
Bulletin. 1887+. Size: 1901-05, 1.

0514. *Rivista di mineralogia e cristallografia italiana*. Padova, 1887+. Edited by R. Panebianco. Size: 1901-05, 2. Ceased 1918.
Bolton 7874.

0515. Geologische Landesanstalt von Elsass-Lothringen. Strassburg. (Publication record dates from 1875)
Mittheilungen. 1888+. Size: 1901-05, 1.
Bolton 7081 (cf. 5005).

0516. *The American geologist*. Minneapolis, 1888+. Edited by W.H. Winchell. Size: 1901-05, 7. Succeeded in 1905 by *Economic geology*.
Bolton 5175. Documentation: DSB--article on N.H. Winchell.

0517. *Eclogae geologicae Helvetiae*. Basel, etc., 1888+. Published by the Société Géologique Suisse. (Includes reports of the Schweizerische Paläontologische Gesellschaft--0497) Size: 1901-05, 1.
Bolton 6047.

0518. Bayerisches Oberbergamt. (Publication record dates from 1861)
Geognostische Jahreshefte. Cassel, later München, 1888+. Size: 1901-05, 1. Ceased 1927.
Bolton 6369.

0519. Service de la Carte Géologique de France et des Topographies Souterraines.
Bulletin. Paris, 1889+. Size: 1901-05, 3.

0520. Geological Survey of New South Wales.
Records. Sydney, 1889+. Size: 1901-05, 1.
Bolton 7648.

0521. Danmarks Geologiske Undersøgelse. [Geological Survey of Denmark]
Various publications, the earliest beginning in 1890. Size (combined): 1901-05, 1.

0522. Geological Society of America. (Founded 1889)
Bulletin. Rochester, N.Y., 1890+. Size: 1901-05, 5.

0523. Iowa Geological Survey.
Annual report. Des Moines, 1892+. Size: 1901-05, 2.

0524. Geological Institution of the University of Upsala.
Bulletin. 1892 (1893)+. Size: 1901-05, 2.

0525. *Journal of geology*. Chicago, 1893+. Published by the Department of Geology, University of Chicago. Edited by T.C. Chamberlin et al. Size: 1901-05, 4.
Bolton 6781.

0526. *Zeitschrift für praktische Geologie*. Berlin, 1893+. Edited by M. Krahmann. Size: 1901-05, 5.
Bolton 8441.

0527. *Rassegna mineraria (metallurgica e chimica)*. [Title varies] Torino, 1894+. Size: 1901-05, 3.

0528. *Rivista italiana di paleontologia*. Bologna, 1895+. Edited by

C. Fornasini and V. Simonelli. Size: 1901-05, 2.
Bolton 7858.

0529. *Ezhegodnik po geologii i mineralogii Rossii. (Annuaire géologique et minéralogique de la Russie)* Warsaw/Novo-Alexandria, 1895+. Size: 1901-05, 1. Ceased 1917.
Bolton 8489.

0530. Società Sismologica Italiana. Roma.
Bollettino. 1895+. Size: 1901-05, 2.

0531. Geological Commission of the Cape of Good Hope.
Annual report. Cape Town, 1896 (1897)+. Size: 1901-05, 1.

0532. Geological Society of South Africa.
Transactions. Johannesburg, 1896+. Size: 1901-05, 2.

0533. *Bibliographia palaeontologica*. Zürich, 1896+. Published by the Concilium Bibliographicum.

0534. *Palaeontographia italica*. Pisa, 1896+. Size: 1901-05, 1.

0535. Geological Survey of the United Kingdom. See also 0472.
Summary of progress. London, 1897+. Size: 1901-05, 2.

0536. Maryland Geological Survey.
Reports. Baltimore, 1897+. Size: 1901-05, 2.

0537. K. Akademie der Wissenschaften in Wien. Erdbeben-Kommission.
Mitteilungen. 1897+. 1901-05, 2.

0538. Vermont Geological Survey.
Report. Burlington, 1898+. Size: 1901-05, 1.

0539. *Pochvovedenie*. [*Pedology*. (Soil science)] St. Petersburg, 1899+. Size: 1901-05, 2.

GEOGRAPHY

The citations counted in the Royal Society's *Catalogue of Scientific Papers* and the *International Catalogue of Scientific Literature*, on which the size estimates of the periodicals are based, were concerned only with physical and mathematical geography. However, most of the periodicals listed below also included much descriptive and economic geography. Consequently their sizes were considerably greater than is indicated by the estimates given.

0540. Société de Géographie. Paris.
Bulletin. 1822+. Size: 1822-63, 2; 1864-73, 2. Cont'd from 1900 as *La géographie: Bulletin de la Société....* Size: 1901-05, 4.
Scudder 1580.

0541. (Royal) Geographical Society of London. See also 0550.
Journal. 1832+. Size: 1832-63, 3; 1864-73, 2. In 1881 it was incorporated in the Society's *Proceedings* (0550).
Scudder 289.

0542. Gesellschaft für Erdkunde zu Berlin. See also 0548 and 0558.

Jährliche Übersicht. 1833+. Cont'd from 1839 as *Monatsbericht.* Size: 1840-53, 1. In 1854 it was incorporated in the *Zeitschrift für allgemeine Erdkunde* (0548).
Scudder 2322.

0543. Geographischer Verein zu Frankfurt a.M. *From 1854* Frankfurter Verein für Geographie und Statistik.
Jahresbericht. 1836/37 (1837)+. Size: 1901-05, 1.
Scudder 2592.

0544. Bombay Geographical Society.
Transactions. [Title varies] 1836/38+. Size: 1846-63, 1. Ceased 1871.
Scudder 3776.

0545. *Archiv für wissenschaftliche Kunde von Russland.* Berlin, 1841+. Edited by G.A. Erman. Size: 1841-60, 1. Ceased 1867.
Scudder 2280. Bolton 598. Documentation: DSB--article on Erman.

0546. United States. Coast Survey. *From 1878* Coast and Geodetic Survey.
Report. 1851+. Size: 1853-63, 1; 1901-05, 1.
Scudder 4352.

0547. American Geographical and Statistical Society. *From 1865* American Geographical Society of New York.
Bulletin. New York, 1852+. Cont'd from 1859 as *Journal.* Size: 1873-74, 2; 1901-05, 1.
Scudder 4168.

0548. *Zeitschrift für allgemeine Erdkunde.* Berlin, 1853+. Published by the Gesellschaft für Erdkunde zu Berlin (0542). Cont'd from 1866 as *Zeitschrift der Gesellschaft....* Size: 1866-73, 1; 1901-05, 3.
Scudder 2322f. Bolton 4835.

0549. Verein für Erdkunde und verwandte Wissenschaften. Darmstadt. (Founded 1845)
Notizblatt. 1854 (1855)+. Size: 1857-63, 1. 1901-05, 1.
Scudder 2492.

0550. (Royal) Geographical Society of London. See also 0541.
Proceedings. 1855/56 (1857)+. Size: 1857-63, 2; 1864-73, 2. Cont'd from 1893 as *The geographical journal.* Size: 1901-05, 9.
Scudder 289.

0551. Justus Perthes' Geographische Anstalt.
Mittheilungen. Gotha, 1855+. Edited by A. Petermann. Size: 1855-63, 2. Cont'd from 1879 as *Petermanns geographische Mittheilungen aus Perthes' geographischer Anstalt.* Size: 1901-05, 7.
Scudder 2668. Bolton 3062 & p. 843.

0552. K.-K. Geographische Gesellschaft in Wein. See also 0571.
Mittheilungen. 1857+. Size: 1857-64, 3; 1901-05, 2.
Scudder 3587.

0553. *Globus: Illustrierte Zeitschrift für Länder- und Völkerkunde.* Hildburghausen, 1861+. Size: 1901-05, 5.
Scudder 2804. Bolton 1993 & p. 756.

0554. Imp. Russkoe Geograficheskoe Obshchestvo. (Société Impériale

Géographique de Russie) St. Petersburg. See also 0585.
Izvestiya (Bulletin). 1865+. Size: 1901-05, 2.
Scudder 3741.

0555. *Geographisches Jahrbuch.* Gotha, 1866+. Edited by E. Behm. Size: 1901-05, 1.
Scudder 2660. Bolton 1890.

0556. Società Geografica Italiana. Roma.
Bollettino. 1868+. Size: 1901-05, 2.
Scudder 1850.

0557. International Geographical Congress. [Also in French, German, and Italian]
[Publications] Anvers, etc, 1872+. Size: 1901-05, 4.
Scudder 11.

0558. Gesellschaft für Erdkunde zu Berlin. See also 0542.
Verhandlungen. 1873+. Size: 1901-05, 1.
Scudder 2322.

0559. Aardrijskundig Genootschap. [Geographical Society] Amsterdam. *Later* K. Nederlandsch Aardrijskundig Genootschap.
Tijdschrift. 1876+. Size: 1901-05, 2.
Scudder 718.

0560. *Geografisk tidsskrift.* Kjøbenhavn, 1877+. Published by the K. Dansk Geografisk Selskab. Size: 1901-05, 1.
Bolton 1884.

0561. *Deutsche geographische Blätter.* Bremen, 1877+. Published by the Geographische Gesellschaft in Bremen. Size: 1901-05, 1.
Bolton 1431.

0562. *Deutsche Rundschau für Geographie und Statistik.* Wien, 1878+. Size: 1901-05, 2. Ceased 1915.
Bolton 1444.

0563. Deutscher Geographentag.
Verhandlungen. Berlin, 1881+. Size: 1901-05, 2. Ceased 1936.

0564. *Scottish geographical magazine.* Edinburgh, 1885+. Published by the Scottish Geographical Society. Size: 1901-05, 3.
Bolton 7990.

0565. *National geographic magazine.* New York, 1888+. Published by the National Geographic Society. Size: 1901-05, 2.
Bolton 7180.

0566. *Annales de géographie.* Paris, 1891+. Size: 1901-05, 2.
Bolton 5235.

0567. *Zemlevyedynie.* [*Geography*] Moskva, 1892+. Size: 1901-05, 1.

0568. *Rivista geografica italiana.* Roma, 1893+. Size: 1901-05, 1.
Bolton 7854.

0569. *Geographische Zeitschrift.* Leipzig, 1895+. Size: 1901-05, 3.
Bolton 6376.

0570. *Zeitschrift für Gewässerkunde.* Leipzig, 1898+. 1901-05, 2.

0571. K.-K. Geographische Gesellschaft in Wien. See also 0552.
Abhandlungen. 1899+. Size: 1901-05, 1.

OTHER EARTH SCIENCES

(a) Geodesy

0572. (Allgemeine) Conferenz der Internationalen Erdmessung. Conférence Géodésique Internationale. (The first meeting was held at Berlin in 1864)
Verhandlungen. *Comptes rendus*. 1864+. Size: 1901-05, 3. Ceased 1913.

0573. K. Preussisches Geodätisches Institut in Potsdam.
Veröffentlichungen. Berlin, 1870+. Size: 1901-05, 1.
Scudder 2371, 2996.

0574. *Zeitschrift für Vermessungswesen*. Stuttgart, 1872+. Size: 1901-05, 5.
Scudder 3245. Bolton 4888.

(b) Meteorology

Titles of a large number of periodicals issued by meteorological and geomagnetic observatories were found (often the one observatory covered both fields) but the great majority of them were not cited in the Royal Society's *Catalogue of Scientific Papers* or the *International Catalogue of Scientific Literature* and so were not included.

0575. Société Météorologique de France. Paris.
Annuaire. 1853 (1855)+. Size: 1901-05, 3.
Scudder 1590.

0576. Scottish Meteorological Society. Edinburgh.
(Quarterly) Report. 1856 (1857)+. Cont'd from 1864 as *Journal*. Size: 1866-73, 4; 1901-05, 1. Ceased 1919.
Scudder 126.

0577. British Meteorological Society. London. (Founded 1850. It was preceded by the Meteorological Society of Great Britain which began in 1838 but lapsed after a few years.) *From 1867* Meteorological Society. *From 1883* Royal Meteorological Society.
Proceedings. 1861+. Size: 1864-73, 3. Cont'd from 1871 as *Quarterly journal*. 1901-05, 3.
Scudder 227.

0578. Collegio Romano. Osservatorio. See also 0332.
Bulletino meteorologico. Roma, 1862+. Edited by A. Secchi. Size: 1862-73, 2. Ceased 1888.
Scudder 2038. Bolton 1174.

0579. *Symons's (monthly) meteorological magazine*. Later *The meteorological magazine*. London, 1866+. Edited by G.J. Symons. Size: 1866-72, 1; 1901-05, 6.
Scudder 464. Bolton 4411.

0580. Österreichische Gesellschaft für Meteorologie.
Zeitschrift. Wien, 1866+. Size: 1866-73, 4. In 1886 it was merged in the *Meteorologische Zeitschrift* (0582).
Scudder 3618. Bolton 8399.

0581. International Meteorological Conference. [Also in French and German] (The first meeting was held at Leipzig in 1872)
Report. 1872+. Size: 1901-05, 1.

0582. *Meteorologische Zeitschrift*. Berlin, 1884+. Published by the Deutsche Meteorologische Gesellschaft (founded 1883). In 1886 it united with the *Zeitschrift der Österreichischen Gesellschaft für Meteorologie* (0580) and cont'd as *Meteorologische Zeitschrift* published at Vienna under the joint sponsorship of the two societies. Size: 1901-05, 11.
Bolton 7022, 8399a. Documentation: DSB--article on J.F. von Hann.

0583. *Das Wetter*. Leipzig, 1884+. Edited by Assmann. Size: 1901-05, 6.
Bolton 8358.

0584. *Monthly weather review*. Washington, 1884+. Published by the Weather Bureau, U.S. Department of Agriculture. Size: 1901-05, 8.

0585. *Meteorologicheskii vestnik*. St. Petersburg, 1892+. Published by the Imp. Russkoe Geograficheskoe Obshchestvo (0554). Size: 1901-05, 3.

(c) Terrestrial Magnetism

See note at the head of the section for Meteorology.

0586. Magnetischer Verein. Göttingen. (Founded 1834)
Resultate aus den Beobachtungen. 1836 (1837)+. Edited by C.F. Gauss and W. Weber. Size: 1837-42, 2. Ceased 1841 (1843).
Scudder 2690. Documentation: Biographies of Gauss.

0587. *Beiträge zur Geophysik*. Stuttgart, 1887+. Edited by G. Gerland. Size: 1901-05, 3.
Bolton 5516.

0588. *Terrestrial magnetism* (later *and atmospheric electricity*). Chicago, 1896+. Edited by L.A. Bauer. Size: 1901-05, 2.
Bolton 8135. Documentation: DSB--article on Bauer.

(d) Oceanography

See also Marine Zoology.

0589. *Wissenschaftliche Meersuntersuchungen*. Hrsg. von der Kommission zur wissenschaftlichen Untersuchungen der deutschen Meere in Kiel, und der biologischen Anstalt auf Helgoland. Berlin, 1871+. Size: 1901-05, 1 (various biological sciences).

0590. K.-K. Hydrographisches Amt. Pola.
Mittheilungen aus dem Gebiete des Seewesens. 1873+. Size: 1901-05, 1.
Scudder 3480. Bolton 3072.

0591. *Hydrographische Mittheilungen*. Berlin, 1873+. Cont'd from 1876 as *Annalen der Hydrographie und maritimen Meteorologie*. Organ des hydrographischen Bureaus und der deutschen Seewarte. Size: 1901-05, 9.
Bolton 2100.

0592. Norddeutsche (*later* Deutsche) Seewarte. Hamburg.
Aus dem Archiv der Deutschen Seewarte. 1878+. Size: 1901-05, 1.
Bolton 5455.

NATURAL HISTORY

Including periodicals covering both botany and zoology, and the field sciences generally.

The distinction between this category and Science in General is rather indefinite: many of the smaller general science periodicals contained only natural history while some of the periodicals listed here contained papers in other sciences, especially in the late nineteenth century.

0593. *Der Naturforscher.* Halle, 1774-1804. Edited by J.E.I. Walch, then by J.C.D. von Schreiber.
See Index. Scudder 2728. Bolton 3263. Documentation: Geus, 1971.

0594. Gesellschaft Naturforschender Freunde. Berlin. (Founded 1773)
Beschäftigungen. 1775+. Cont'd from 1780 as *Schriften.* Cont'd from 1795 as *Neue Schriften.* Size: 1795-1805, 1. Cont'd from 1807 as *Magazin.* Size: 1807-18, 2. Cont'd from 1819 as *Verhandlungen.* Size: 1819-29, 1. Cont'd from 1836 as *Mittheilungen.* Size: 1836-39, 2. Ceased 1838 (1839); for its successor see 0639.
See Index. Scudder 2292, 2323. Bolton 909. Documentation: Geus, 1971.

0595. Linnean Society of London. (Founded 1788) See also 0613, 0707, and 0767.
Transactions. 1788 (1791)+. Size: 1791-1863, 1; 1864-73, 1. From 1875 it was issued in two sections: *Transactions. Botany* and *Transactions. Zoology.* Size (Botany): 1901-05, no references found; (Zoology): 1901-05, 1.
See Index. Scudder 333.

0596. Société Impériale des Naturalistes de Moscou. See also 0606.
Mémoires. 1806+. Size: 1806-23, 1. Ceased in 1823 but resumed in 1829 as *Nouveaux mémoires.* Size: 1829-55, 1. Cont'd past 1900.
Scudder 3693.

0597. Wernerian Natural History Society. Edinburgh.
Memoirs. 1808 (1811)+. Size: 1808-37, 1. Ceased 1838 (1839).
Scudder 129. Documentation: DSB--article on R. Jameson.

0598. Academy of Natural Sciences of Philadelphia. (Founded 1812) See also 0616
Journal. 1817 (1818)+. Suspended 1839 (1842) - 1847. Size: 1817-39, 2; 1847-62, 2; 1901-05, 1. Ceased 1918.
Scudder 4232.

0599. Lyceum of Natural History of New York. *From 1877* New York Academy of Sciences (0230).

Annals. 1823 (1824)+. Size: 1824-63, 1; 1864-73, 2. For continuation after 1876 see 0230.
Scudder 4199.

0600. Société d'Histoire Naturelle de Paris. (Founded 1791)
Mémoires. 1823+. Size: 1823-34, 2. Ceased 1834.
See Index. Scudder 1583.

0601. Société Linnéenne du Calvados (*from 1826* de Normandie). See also 0637.
Mémoires. Caen/Paris, 1824 (1825)+. Size: 1826-62, 1; 1864-69, 1. Cont'd past 1900.
Scudder 1101, 1102.

0602. *Annales des sciences naturelles*. Paris, 1824+. Edited by J.V. Audouin, A.T. Brongniart, and J.B. Dumas. Size: 1824-33, 11. Cont'd from 1834 in two sections: *Botanique* (0689) and *Zoologie* (0758).
See Index. Scudder 1308. Bolton 377 & p. 629.

0603. Société Linnéenne de Bordeaux.
Bulletin. 1826+. Cont'd from 1830 as *Actes*. Size: 1830-54, 1; 1901-05, 4 (zoology and botany approximately equal).
Scudder 1071.

0604. Société Linnéenne de Lyon.
Annales. 1826/34 (1836)+. Size: 1836-52, 1; 1852-57, 6; 1864-73, 5; 1901-05, 1.
Scudder 1195.

0605. *The magazine of natural history*. London, 1828 (1829)+. Edited by J.C. Loudon until 1837, then by E. Charlesworth. Size: 1829-36, 11; 1837-40, 9. In 1840 it united with *Annals of natural history* (0614) to form *The annals and magazine of natural history* (0614).
Scudder 346. Bolton 2795. Documentation: Sheets-Pyenson, 1981a.

0606. Société Impériale des Naturalistes de Moscou. See also 0596.
Bulletin. 1829+. Size: 1829-36, 4; 1837-63, 2; 1864-73, 5; 1901-05, 3 (about one-third mineralogy, one-third other earth sciences, and one-third various biological sciences).
Scudder 3693.

0607. Société d'Histoire Naturelle (*from 1835* Société du Museum d'Histoire Naturelle; *from 1858* Société des Sciences Naturelles) de Strasbourg. (*Cont'd after 1870 as* Société des Sciences de Nancy--0222)
Mémoires. 1830+. Size: 1830-62, 1. Ceased 1870.
Scudder 1584, 1591, 3238.

0608. The Natural History Society of Northumberland, Durham and Newcastle-upon-Tyne. (Founded 1829)
Transactions. Newcastle, 1831+. Suspended between 1838 and 1865. Size: 1867-74, 4; 1901-05, 1.
Scudder 506.

0609. Berwickshire Naturalists' Club. (Founded 1831)
History (sometimes *Proceedings*). Berwick-upon-Tweed, 1831 (1834)+. Size: 1834-63, 1; 1864-73, 1. Cont'd past 1900.
Scudder 42.

0610. Boston Society of Natural History. See also 0617.
Boston journal of natural history. 1834+. Size: 1834-63, 1. Cont'd from 1866 as *Mémoires*. Size: 1866-73, 1. Cont'd past 1900.
Scudder 3991. Bolton 1016.

0611. *Naturhistorisk tidsskrift*. Kjøbenhavn, 1837+. Edited by H. Krøyer. Size: 1837-42, 4; 1844-49, 2. Suspended 1850-60. Size: 1861-73, 1. Ceased 1884.
Scudder 632. Bolton 3272.

0612. Verein für Naturkunde zu Cassel. (Founded 1836)
Bericht (or *Jahresbericht*). 1837+. Size: 1837-62, 1. Cont'd from 1894 as *Abhandlungen und Bericht*. Size: 1901-05, 1.
Scudder 2474.

0613. Linnean Society of London. See also 0595.
Proceedings. 1838/48 (1849)+. Size: 1838-55, 2; 1864-73, 1; 1901-05, 1.
Scudder 333.

0614. *Annals of natural history; or, Magazine of zoology, botany and geology*. London, 1838-40. Edited by W. Jardine et al. (Richard Taylor, the publisher, was instrumental in its establishment) In 1841 it united with *The magazine of natural history* (0605) and cont'd as *Annals and magazine of natural history, including zoology, botany and geology*. Edited by W. Jardine et al. Size: 1838-47, 7; 1848-57, 6; 1858-63, 11; 1864-73, 11; 1901-05, 11 (zoology 95%, botany 2.5%, palaeontology 2.5%)
See Index. Scudder 192. Bolton 2800 & p. 825. Documentation: Sheets-Pyenson, 1981a.

0615. *Annali* (later *Nuovi annali*) *delle scienze naturali*. Bologna, 1838+. Edited by A. Alessandrini et al. Size: 1838-49, 4; 1850-54, 2. Ceased 1854.
Scudder 1802. Bolton 400.

0616. Academy of Natural Sciences of Philadelphia. See also 0598.
Proceedings. 1841 (1843)+. Size: 1841-63, 5; 1864-73, 4; 1901-05, 7 (zoology 84%, other 16%).
Scudder 4232.

0617. Boston Society of Natural History. See also 0610.
Proceedings. 1841 (1844)+. Size: 1841-63, 3; 1864-73, 4; 1901-05, 1.
Scudder 3991.

0618. Société Vaudoise des Sciences Naturelles. Lausanne.
Bulletin. 1842+. Size: 1842-63, 2; 1864-73, 2; 1901-05, 4 (most sciences, chiefly zoology).
Scudder 2173.

0619. Société d'Histoire Naturelle du Département de la Moselle. Metz. (Founded 1835)
Mémoires. 1843+. From 1844 cont'd as *Bulletin*. Size: 1843-60, 1; 1901-05, 2.
Scudder 3098.

0620. Société des Sciences Naturelles de Neuchâtel.

Bulletin. 1844 (1846)+. 1844-63, 1; 1864-73, 1; 1901-05, 1.
Scudder 2175.

0621\. Naturhistorischer Verein der Preussischen Rheinlande (*later* und Westphalens). Bonn. (Founded 1843)
(a) *Verhandlungen*. 1844+. Size: 1844-63, 3; 1864-73, 1; 1901-05, 2.
(b) *Correspondenzblatt*. 1844+. Size: 1844-63, 1; 1864-73, 2.
(c) *Sitzungsberichte*. 1844(?)+. Size: 1856-63, 2.
The three periodicals were bound together.
Scudder 2411.

0622\. *Allgemeine deutsche naturhistorische Zeitung*. Dresden, 1846+. Published by the Gesellschaft 'Isis' in Dresden (0173). Size: 1846-57, 3. Ceased 1857.
Scudder 2505. Bolton 92.

0623\. Zoologisch-mineralogischer Verein (*from 1883* Naturwissenschaftlicher Verein) in Regensburg. (Founded 1846)
Correspondenz-Blatt. 1857+. Cont'd from 1887 as *Berichte*. Size: 1901-05, 2 (mostly the geological sciences).
Scudder 3198.

0624\. Cotteswold Naturalists' Field Club. Gloucester. (Founded 1846)
Proceedings. 1847 (1853)+. Size: 1901-05, 1.
Scudder 143.

0625\. Verein der Freunde der Naturgeschichte in Mecklenburg. Güstrow.
Archiv. Neubrandenburg, 1847+. Size: 1847-57, 2; 1901-05, 2.
Scudder 3138.

0626\. Naturhistorischer Verein in Augsburg. (Founded 1846) *From 1887* Naturwissenschaftlicher Verein für Schwaben und Neuburg)
Bericht. 1848+. Size: 1848-63, 1; 1864-73, 1; 1901-05, 1.
Scudder 2250.

0627\. Natural History of Dublin.
Proceedings. 1849+. Size: 1849-63, 1. Ceased 1865 (1871).
Scudder 98; cf. 82.

0628\. Naturhistorisk Forening i Kjøbenhavn.
Videnskabelige meddelelser. [*Scientific communications*] 1849+. Size: 1849-59, 1; 1860-64, 2; 1901-05, 4 (zoology 67%, botany 20%, geological sciences 13%).
Scudder 631. Bolton 4688 & p. 976.

0629\. Naturhistorische Gesellschaft. Hannover. (Founded 1797)
Jahresbericht. 1850+. Size: 1859-63, 1; 1901-05, 1.
Scudder 2789, 2784.

0630 *Natuurkundig tijdschrift voor Nederlandsch-Indië*. Batavia, 1850+. Published by the Natuurkundige Vereeniging in Nederlandsch-Indië. Size: 1850-63, 11; 1864-73, 2; 1901-05, 1.
Scudder 3772. Bolton 3283 & p. 857.

0631\. *Lotos: Zeitschrift für Naturwissenschaften*. Prag, 1851+. Published by the Naturhistorischer Verein 'Lotos' in Prag. Size: 1851-63, 2; 1901-05, 1.
Scudder 3504. Bolton 2716 & p. 821.

0632\. (a) *The naturalist*. London, 1851+. Edited by B.R. Morris.

Cont'd from 1860 as *The magazine of natural history and Naturalist.* Cont'd as (b).
(b) *The naturalist: Journal of the West Riding Consolidated Naturalists' Society.* [Sub-title varies] York, 1864+. Suspended 1867-75. Size: 1901-05, 9 (botany 67%, zoology 23%, geological sciences 10%).
Scudder 564. Bolton 3240, 3244, 4818.

0633. (K.-K.) Zoologisch-Botanischer Verein (*from 1858* Gesellschaft) in Wien.
Verhandlungen. 1851 (1852)+. Size: 1852-63, 4; 1864-73, 5; 1901-05, 4 (zoology 68%, botany 32%).
Scudder 3648.

0634. *The natural history review.* Dublin, 1853 (1854)+. Included the transactions of several Irish natural history societies. Size: 1854-60, 3; 1861-65, 2. Ceased 1865.
Scudder 97. Bolton 3238.

0635. Senckenbergische Naturforschende Gesellschaft. Frankfurt a.M. (Founded in 1817 and named after J.C. Senckenberg who established the Frankfurt museum) See also 0643.
Abhandlungen. 1854+. Size: 1854-63, 1; 1864-73, 2; 1901-05, 2.
Scudder 2607.

0636. Società Geologica residente in Milano. *From 1860* Società Italiana di Scienze Naturali.
Atti. Milano, 1855/59 (1859)+. Size: 1860-73, 2; 1901-05, 2.
Scudder 1913, 1915.

0637. Société Linnéenne de Normandie. See also 0601.
Bulletin. Caen, 1855+. Size: 1856-64, 1; 1865-68, 1; 1901-05, 1.
Scudder 1102.

0638. *The Canadian naturalist and geologist.* Montreal, 1856 (1857)+. Edited by E. Billings. Published (from Vol. 2) by the Natural History Society of Montreal. Size: 1856-63, 1; 1864-73, 2. Cont'd from 1869 as *The Canadian naturalist and quarterly journal of science.* Cont'd from 1885 as *The Canadian record of science.* Ceased 1916.
Scudder 4375. Bolton 1202.

0639. Gesellschaft Naturforschender Freunde. Berlin. See also 0594.
Sitzungsberichte. 1860 (1862)+. Size: 1865-70, 1; 1901-05, 4 (various biological sciences, chiefly zoology).
Scudder 2323.

0640. Bristol Naturalists' Society.
Proceedings (or *Annual report).* 1862 (1863)+. Size: 1901-05, 1.
Scudder 53.

0641. Naturforschender Verein in Brünn.
Verhandlungen. 1862 (1863)+. Size: 1862-73, 2; 1901-05, 2.
See Index. Scudder 3357.

..... *Hardwicke's science gossip.* 1865+. Main entry 0192.

0642. *The American naturalist.* Salem, Mass., 1867/68 (1868)+. Edited by A.S. Packard et al. Published initially by the Essex Institute and from Vol. 2 by the Peabody Academy of Science. Size: 1868-73, 10; 1901-05, 7 (zoology 75%, botany 11%, palae-

ontology 6%, other 8%).
Scudder 4302, 4303. Bolton 237.

0643. Senckenbergische Naturforschende Gesellschaft. Frankfurt a.M. See also 0635.
Bericht. 1868 (1869)+. Size: 1901-05, 2.
Scudder 2607.

0644. Norfolk and Norwich Naturalists' Society.
Transactions. Norwich, 1869 (1870)+. Size: 1901-05, 2.
Scudder 513.

0645. Croydon Microscopical Club. (Founded 1870) *From 1902* Croydon Natural History and Scientific Society.
Report. 1870 (1871)+. Size: 1901-05, 1 (geological sciences and botany).
Scudder 76.

0646. *Feuille des jeunes naturalistes*. Paris, 1870+. Edited by A. Dollfus. Size: 1901-05, 5 (zoology 85%, botany 14%). Ceased 1926.
Scudder 2497. Bolton 1692.

0647. (a) *The Scottish naturalist*. Perth, 1870+. Published by the Perthshire Society of Natural Science. Size: 1870-74, 4. Succeeded by (b).
(b) *Annals of Scottish natural history*. Edinburgh, 1892+. Size: 1901-05, 6 (botany and zoology approximately equal).
Scudder 523. Bolton 4309, 5273.

0648. Belfast Natural History and Philosophical Society.
Report (later *Proceedings*). 1871/72+. Size: 1901-05, 1.
Scudder 38.

0649. Société d'Etude des Sciences Naturelles. Nîmes.
Bulletin. 1873+. Size: 1901-05, 1.
Scudder 1261.

0650. Sällskapets (*from 1875* Societas) pro Fauna et Flora Fennica. Helsingfors. (Publication record dates from 1848) See also 0653.
Acta. 1875+. Size: 1901-05, 2.
Scudder 3664, 3665.

0651. Società Toscana di Scienze Naturali. Pisa.
Atti. 1875+. Size: 1901-05, 2.
Scudder 2014.

0652. Linnean Society of New South Wales. Sydney.
Proceedings. 1875/76 (1877)+. Size: 1901-05, 4 (zoology 60%, botany 30%, other 10%)
Scudder 3856.

0653. Sällskapets (*from 1875* Societas) pro Fauna et Flora Fennica. Helsingfors. See also 0650.
Meddelanden. 1876+. Size: 2.
Scudder 3664, 3665.

0654. *Természetrajzi füzetek*. [*Natural history bulletin*] Budapest, 1877+. Published by the Magyar Nemzeti Múzeum [Hungarian National Museum] Size: 1901, 5. Ceased 1902.
Bolton 4469.

0655. Cincinnati Society of Natural History.
Journal. 1878+. Size: 1901-05, 1.

0656. Société des Sciences Naturelles de Saône-et-Loire. Chalon-sur-Saône.
Bulletin. 1878+. Size: 1901-05, 1. Ceased 1924.

0657. *Guide du naturaliste. Revue bibliographique des sciences naturelles.* Paris, 1879+. Edited by A. Bouvier.
Bolton 2015.

0658. Hertfordshire Natural History Society and Field Club.
Transactions. Hertford, 1879+. Size: 1901-05, 1.

0659. Société Fribourgeoise des Sciences Naturelles. (Naturforschende Gesellschaft in Freiburg)
Bulletin. 1879/80+. Size: 1901-05, 1.

0660. *Naturae novitates. Bibliographie neuer Erscheinungen aller Länder auf dem Gebiete der Naturgeschichte.* Berlin, 1879+.

0661. *Bibliographie des sciences naturelles.* Paris, 1880+.
Bolton 5538.

0662. Northamptonshire Natural History Society and Field Club.
Journal. Northampton, 1880+. Size: 1901-05, 1.

0663. Edinburgh Naturalists' Field Club.
Transactions. 1881/86+. Size: 1901-05, 1. Ceased 1915.

0664. Société d'Etude des Sciences Naturelles d'Elbeuf.
Bulletin. 1881/82+. Size: 1901-05, 1.

0665. *Bollettino del naturalista (collettore).* Siena, 1881+. Cont'd from 1889 as *Rivista italiana di scienze naturali.* Size: 1901-05, 2. Ceased 1910.
Bolton 1163 (& p. 669), 7859.

0666. *The Victorian naturalist.* Melbourne, 1884/5+. Published by the Field Naturalists' Club of Victoria. Size: 1901-05, 4 (zoology 66%, botany 20%, other 14%)
Bolton 8285.

0667. Bombay Natural History Society.
Journal. 1886+. Size: 1901-05, 6 (mostly zoology).

0668. Società di Naturalisti in Napoli.
Bollettino. 1887+. Size: 1.

0669. *The Essex naturalist.* Buckhurst Hill, 1887+. Published by the Essex Field Club. Size: 1901-05, 1.
Bolton 6138.

0670. Société des Sciences Naturelles de l'Ouest de la France. Nantes.
Bulletin. 1891+. Size: 1901-05, 1.

0671. *Natur und Haus.* Berlin, 1892+. Size: 1901-05, 9 (mostly zoology with some botany).
Bolton 7183.

0672. *The Irish naturalist.* Dublin, 1892+. Published by several Irish societies jointly. Size: 1901-05, 7 (botany 50%, zoology 35%, geological sciences 12%, general biology 3%).
Bolton 6639.

0673. *Revista chilena de historia natural*. Valparaiso, 1897+. Published by the Museo de Historia Natural. Size: 1901-05, 1.

MICROSCOPY

Microscopic anatomy is included in Experimental Biology.

0674. (Royal) Microscopical Society. London. (Founded 1839)
Transactions. 1844+. Size: 1844-52, 1; 1853-68, 1. Cont'd from 1869 as *Monthly microscopical journal: Transactions*. Size: 1869-73, 9. Cont'd from 1878 as *Journal*. Size: 1901-05, 1.
Scudder 376. Bolton 3174.

0675. *Quarterly journal of microscopical science*. London, 1853+. Edited by E. Lankester et al. Size: 1853-63, 3; 1864-73, 3; 1901-05, 2.
See Index. Scudder 419. Bolton 3826.

0676. Quekett Microscopical Club. London. (Founded 1865; named after J.T. Quekett (1815-1861), a prominent British microscopist)
Journal. 1868+. Size: 1868-73, 1; 1901-05, 2.
Scudder 423.

0677. National Microscopical Congress. *From 1879* American Society of Microscopists. *From 1891* American Microscopical Society.
Proceedings, later *Transactions*. Lincoln, Nebr., 1878+. Size: 1901-05, 1.

0678. Manchester Microscopical Society.
Annual report, later *and Transactions*. 1880+. Size: 1901-05, 1. Ceased 1935.

0679. *Zeitschrift für wissenschaftliche Mikroskopie und für mikroskopische Technik*. Braunschweig, 1884+. Edited by W.J. Behrens. Size: 1901-05, 7.
Bolton 8451. Documentation: Freund, 1964.

0680. *Zeitschrift für angewandte Mikroskopie*. Leipzig, 1895+. Edited by G. Marpmann. Size: 1901 05, 2.
Bolton 8404.

BOTANY

Including some horticultural periodicals.

Much botany is also included in the periodicals listed under Natural History.

0681. *Botanisches Magazin* (sometimes *Magazin für die Botanik*). Zürich, 1787+. Edited by J.J. Römer and P. Usteri. Cont'd from 1791 as *Annalen der Botanik*. Ceased 1800.
Scudder 2195, 2191. Bolton 2778. Documentation: Kronick, 1976, p. 264.

0682. *The botanical magazine, or flower garden displayed.* London, 1787+. By W. Curtis. Cont'd from 1801 as *Curtis's botanical magazine.* Edited from 1826 by W.J. Hooker and later by J.D. Hooker; as a result it became closely associated with the Royal Botanic Gardens at Kew.
Bolton 1025. Documentation: Blunt, 1950; Hunkin, 1946.
Not a periodical in the usual sense but a serial containing detailed, accurate, and often beautiful illustrations of plants. Very successful from its foundation to the present day.

0683. *The botanist's repository for new and rare plants.* London, 1797+. Edited by H.C. Andrews. Ceased 1815.
Bolton 1049.

0684. (Royal) Horticultural Society. London. See also 0700.
Transactions. 1810 (1812)+. Size: 1812-30, 1; 1831-48, 1. Ceased 1848.

0685. *Flora; oder, Botanische Zeitung.* Regensburg, 1818+. Published by the K. Bayerische Botanische Gesellschaft in Regensburg (founded 1790). Size: 1818-63, 4; 1864-73, 2; 1901-05, 4.
See Index. Scudder 3191. Bolton 1706.

0686. *Årsberättelse om botaniska arbeten och upptäcker.* Stockholm, 1820 (1821) - 1852 (1857). Published by the K. Vetenskaps Akademien (see 074) and edited by J.E. Wikström.
——— [Trans.] *Jahresbericht ... über die Fortschritte der Botanik.* Breslau, 1820 (1834) - 1843 (1848). Trans. by C.T. Beilschmied.
Scudder 693f, 2460. Bolton 652, 2282.

0687. *Linnaea: Ein Journal für die Botanik in ihrem ganzen Umfange.* Berlin, 1826+. Edited by D.F.L. von Schlechtendal. Size: 1826-63, 2; 1864-73, 1. Ceased 1882.
Scudder 2342. Bolton 2696 & p. 819.

0688. (a) *The botanical miscellany.* London, 1830+. Size: 1830-33, 1. Cont'd as (b).
(b) *The journal of botany.* Ib., 1834+. Size: 1834-42, 3. Cont'd as (c).
(c) *The London journal of botany.* Ib., 1842+. Size: 1842-48, 3. Cont'd as (d).
(d) *Hooker's journal of botany and Kew Garden miscellany.* Ib., 1849+. Size: 1849-57, 2. Ceased 1857.
All four journals were edited by W.J. Hooker.
Scudder 221, 321, 340, 301. Bolton 1026.

0689. *Annales des sciences naturelles: Botanique.* Paris, 1834+. (For preceding series see 0602) Size: 1834-43, 5; 1844-53, 3; 1854-63, 2; 1864-73, 3; 1901-05, 1.
Scudder 1308. Bolton 377b & p. 629.

0690. (a) *Jahresbericht über die Resultate der Arbeiten im Felde der physiologischen Botanik.* Berlin, 1834-37 (1837-39). By F.J.F. Meyen. Separate publication of one of the many annual reports included in the *Archiv für Naturgeschichte* (0759). Cont'd as (b).
(b) *Jahresbericht über die Arbeiten für physiologische Botanik.* Ib., 1838 (1840)+. By H.F. Link. Ceased 1846 (1849).
Scudder 2336. Bolton 2302.

0691. *Icones plantarum (*later *Hooker's icones plantarum).* London, 1836+. Edited by W.J. Hooker and from 1867 to 1890 by J.D. Hooker.

0692. Botanical Society of Edinburgh. (Founded 1836)
Annual report and proceedings. 1836 (1837)+. Cont'd from 1844 as *Transactions (*later *and proceedings).* Size: 1844-63, 2; 1864-73, 4; 1901-05, 2.
Scudder 108.

0693. *Botaniska notiser.* Lund, 1839. (Sometimes *Nya botaniska notiser*) Edited by A.E. Lindblom. Size: 1863-72, 3; 1901-05, 3.
Scudder 671. Bolton 1045.

0694. *Bericht über die Leistungen in der Pflanzengeographie und systematischen Botanik.* Berlin, 1840 (1843)+. Edited by A. Grisebach. Separate publication of one of the many annual reports included in the *Archiv für Naturgeschichte* (0759). Ceased 1854 (1857).
Scudder 2288. Bolton 888.

0695. *The gardener's chronicle.* London, 1841+. Edited by J. Lindley. Size: 1841-63, 1; 1864-73, 2; 1901-05, 10.
Bolton 1810.

0696. *The phytologist. A popular botanical miscellany.* London, 1842 (1844)+. Edited by G. Luxford and E. Newman. Size: 1844-63, 2. Ceased 1863.
Scudder 415. Bolton 3667.

0697. *Repertorium botanicae systematicae.* Leipzig, 1842+. Edited by W.G. Walpers. Cont'd from 1848 as *Annales botanicae systematicae.* Ceased 1868.
Scudder 2998, 2887.

0698. *Botanische Zeitung.* Berlin, etc., 1843+. Edited by H. Mohl and D.F.L. von Schlechtendal. Size: 1843-63, 6; 1864-73, 5; 1901-05, 3. Ceased 1910.
Scudder 2708. Bolton 1034.
"One of the most famous periodicals of modern botany."--DSB (article on Mohl)

0699. *Giornale botanico italiano.* Firenze, 1844+. Edited by the Sezione Botanica dei Congressi Scientifici Italiani. Ceased in 1852 but revived in 1869 as *Nuovo giornale botanico italiano.* Edited by T. Caruel. Later published by the Società Botanica Italiana. Size: 1869-73, 1; 1901-05, 2.
Scudder 1834, 1843. Bolton 1934, 1935 & p. 754.

0700. (Royal) Horticultural Society. London. See also 0684.
Journal. 1846+. Size: 1846-55, 2. Suspended 1855-66. 1901-05, 3.

0701. *Nederlandsch kruidkundig archief.* Leiden, 1846+. Edited by W.H. de Vriese et al. From 1871 published by the Nederlandsche Botanische Vereeniging. Size: 1846-73, 1; 1901-05, 1.
Scudder 878, 893. Bolton 3307.

0702. *Österreichisches botanisches Wochenblatt.* Wien, 1851+. Edited by A. Skofitz. Size: 1851-57, 11. Cont'd from 1858 as *Österreichische botanische Zeitschrift.* Size; 1858-63, 4; 1864-72,

3; 1901-05, 3.
Scudder 3616, 3617. Bolton 3451.

0703. *Hedwigia. Ein Notizblatt für kryptogamische Studien.* Dresden, 1852+. Edited by L. Rabenhorst. Size: 1865-72, 1; 1901-05, 3.
Scudder 2507. Bolton 2042 & p. 763.

0704. *Die Gartenflora.* Erlangen, 1852+. Published by the Deutsche Gartenbau-Gesellschaft. Size: 1901-05, 7.
Bolton 4247b,c.

0705. *Bonplandia. Zeitschrift für die gesammte Botanik.* Hannover, 1853+. Published by the K. Leopoldinisch-Carolinische Akademie der Naturforscher (03) until 1859, then edited by W. and B. Seemann. Size: 1856-62, 3. Ceased 1862.
Scudder 2783, 2779. Bolton 1012.

0706. Société Botanique de France. Paris.
Bulletin. 1854+. Size: 1854-63, 4; 1864-73, 6; 1901-05, 5.
Scudder 1565.

0707. Linnean Society of London. See also 0595.
Journal (of the proceedings). Botany. 1855 (1857)+. Size: 1857-63, 2; 1864-73, 2; 1901-05, 2.
Scudder 333.

0708. *Jahrbücher für wissenschaftliche Botanik.* Berlin, 1858. Edited by N. Pringsheim until his death in 1894. Size: 1864-73, 2; 1901-05, 4.
Scudder 2332. Bolton 2280.

0709. Botanischer Verein für die Provinz Brandenburg. Berlin.
Verhandlungen. 1859+. Size: 1859-73, 1; 1901-05, 3.
Scudder 2303.

0710. *Adansonia. Recueil périodique d'observations botaniques.* Paris, 1860+. Edited by H. Baillon. Size: 1860-73, 4. Ceased 1879.
Scudder 1280. Bolton 33.

0711. Preussischer Botanischer Verein zu Königsberg.
Jahresbericht. 1862+. Size: 1901-05, 1.

0712. Société Royale de Botanique de Belgique. Bruxelles.
Bulletin. 1862+. Size: 1862-73, 2. Cont'd past 1900.
Scudder 958.

0713. *The journal of botany, British and foreign.* London, 1863+. Edited by B. Seemann; later (from 1880?) by J. Britten. Size: 1864-73, 8; 1901-05, 11.
See Index. Scudder 322. Bolton 2503.

0714. International botanical congress. [Also in French and Italian] (The first meeting was held at Brussels in 1864)
Actes. 1864+.
Scudder 8, 9, 14, 16.

0715. *Botanisk tidsskrift.* Kjøbenhavn, 1866+. Published by the Botaniske Forening i Kjøbenhavn. From 1875 also entitled *Journal de botanique.* Size: 1866-71, 2; 1901-05, 4.
Scudder 611. Bolton 1044.

0716. Torrey Botanical Club. (Formerly New York Botanical Club) Named after John Torrey (1796-1873), a prominent American

botanist.
Bulletin. New York, 1870+. Size: 1901-05, 6.
Scudder 4223.

0717. Société Botanique de Lyon.
Annales. 1871+. Size: 1901-05, 1.
Scudder 1189.

0718. *The garden*. London, 1871+. Size: 1901-05, 3.
Bolton 1802.

0719. (R.) Università di Pavia. Laboratorio di Botanica Crittogamica (*later* Istituto Botanico).
Archivio, later *Atti*. 1871/73 (1874)+. Size: 1901-05, 2.
Scudder 1996.

0720. *Grevillea. A record of cryptogamic botany and its literature*. London, 1872 (1873)+. Edited by M.C. Cooke. Size: 1873-74, 1. Ceased 1894.
Scudder 298. Bolton 2012.

0721. *Botanischer Jahresbericht: Systematisch geordnetes Repertorium der botanischen Literatur aller Länder*. Berlin, 1873 (1874)+. Edited by L. Just. Cont'd from 1883 as *Just's botanischer Jahresbericht....*
Scudder 2302. Bolton 1035.

0722. *Revue bryologique*. Caen, 1874+. Edited by T. Husnot. Size: 1901-05, 2.
Bolton 4017.

0723. *The botanical bulletin*. Hanover, Ind., 1875+. Edited by J.M. Coulter. Cont'd from 1876 as *The botanical gazette*. Size: 1901-05, 6.
Scudder 4080. Bolton 1022.

0724. Jardin Botanique de Buitenzorg. [Java]
Annales. Batavia, 1876+. From 1882 edited by M. Treub. Size: 1901-05, 1.
Bolton 391. Documentation: DSB--article on Treub.

0725. *Botanisches Centralblatt*. Cassel, 1880+. Edited by O. Uhlworm. Partly an abstracting journal. Size: 1901-05, 7.
Bolton 1037 & p. 671.

0726. *Botanische Jahrbücher für Systematik, Pflanzengeschichte und Pflanzengeographie*. Leipzig, 1881+. Edited by A. Engler. Size: 1901-05, 5.
See Index. Bolton 1032. Documentation: DSB (Vol. XV)--article on Engler.

0727. *Deutsche botanische Monatsschrift*. Sondershausen, 1883+. Edited by G. Leimbach. Size: 1901-05, 3. Ceased 1912.
Bolton 5963.

0728. Deutsche Botanische Gesellschaft. Berlin. (Founded 1882)
Berichte. 1883+. Size: 1901-05, 10.

0729. *Journal of mycology*. Columbus, Ohio, 1885+. Suspended 1894-1902. Size: 1902-05, 2.
Bolton 6785.

0730. *Malpighia: Rassegna mensuale di botanica*. Messina, etc., 1886+. Edited by O. Penzig et al. Size: 1901-05, 2.
Bolton 6926.

0731. Royal Botanic Gardens. Kew.
Bulletin (of miscellaneous information). London, 1887+.

0732. *Annals of botany*. Oxford/London, 1887+. Edited by I.B. Balfour, S.H. Vines, and W.G. Farlow. Size: 1901-05, 4.
Bolton 5269.

0733. *Journal de botanique*. Paris, 1887+. Edited by L. Morot. Size: 1901-05, 2. Ceased 1913.
Bolton 6727.

0734. *The botanical magazine*. Tokyo, 1887+. Published by the Tokyo Botanical Society. Size: 1901-05, 2.
Bolton 5625.

0735. *Revue générale de botanique*. Paris, 1889+. Edited by G. Bonnier. Size: 1901-05, 3.
Bolton 7747.

0736. *Le monde des plantes*. Paris, 1891+. Cont'd later as *Bulletin de géographie botanique*. Published by the Académie Internationale de Géographie Botanique. Size: 1901-05, 3. Ceased 1919.
Bolton 7121.

0737. Botanischer Verein für Gesamtthüringen. Weimar. (Founded 1882) *Later* Thüringisch-Botanischer Verein.
Mitteilungen. 1891+. Size: 1901-05, 1.

0738. *Monatsschrift für Kakteenkunde*. Neudamm, 1891+. Size: 1901-05, 4.
Bolton 7115.

0739. Bayerische Botanische Gesellschaft zur Erforschung der Heimischen Flora. München. (Founded 1889) See also 0740.
Bericht. 1891+. Size: 1901-05, 1.

0740. Bayerische Botanische Gesellschaft zur Erforschung der Heimischen Flora. See also 0739.
Mitteilungen. 1892+. Size: 1901-05, 3.

0741. Società Botanica Italiana.
Bullettino. Firenze, 1892+. Size: 1901-05, 4.

0742. Deutsche Dendrologische Gesellschaft.
Mitteilungen. Bonn, etc., 1892+. Size: 1901-05, 2.

0743. *The Linnaean fern bulletin*. Binghampton, N.Y., 1893+. Cont'd from 1896 as *The fern bulletin*. Published by the American Fern Society. Size: 1901-05, 5.

0744. Herbier Boissier. Chambésy.
Bulletin. Genève, etc., 1893+. Size: 1901-05, 3.

0745. *Minnesota botanical studies*. Minneapolis, 1894+. Published by the Geological and Natural History Survey of Minnesota. Size: 1901-05, 1.

0746. K. Botanischer Garten und Museum zu Berlin.
Notizblatt. Leipzig, 1895+. Size: 1901-05, 1.
Bolton 7275.

0747. *Allgemeine botanische Zeitschrift für Systematik, Floristik, Pflanzengeographie, etc.* Karlsruhe, 1895+. Edited by A. Kneucker. Size: 1901-05, 5. Ceased 1927.
Bolton 5093.

0748. Conservatoire et Jardin Botaniques de Genève.
Annuaire. 1897+. Size: 1901-05, 2.

0749. *The plant world. A monthly journal of general botany.* Binghamton, N.Y., 1897+. Size: 1901-05, 4. Cont'd from 1920 as *Ecology*.

0750. *The bryologist.* Brooklyn, N.Y., etc., 1898+. Size: 1901-05, 3.

0751. *Rhodora.* Boston, 1899+. Published by the New England Botanical Club. Size: 1901-05, 9.

0752. International Conference on (Plant-breeding and) Hybridisation. [Also in French and German]
[*Proceedings*] Chiswick, etc., 1899+.

ZOOLOGY

Much zoology is also included in the periodicals listed under Natural History.

(a) General Zoology

0753. Muséum National d'Histoire Naturelle. Paris. See also 0800.
Annales. 1802+. Size: 1802-13, 10. Cont'd from 1815 as *Mémoires.* Size: 1815-32, 4. Cont'd from 1832 as *Nouvelles annales.* Size: 1832-35, 3. Cont'd from 1839 as *Archives.* Size: 1839-61, 1. Cont'd from 1865 as *Nouvelles archives.* Size: 1865-73, 1; 1901-05, 1.
See Index. Scudder 1463. Bolton 394, 1136.

0754. *Isis; oder, Encyclopädische Zeitung* (from 1828 *vorzüglich für Naturgeschichte, vergleichende Anatomie und Physiologie.* Jena, 1817+. Edited by L. Oken. Size: 1817-48, 6. Ceased 1848.
See Index. Scudder 2819. Bolton 2229. Documentation: DSB--article on Oken.

0755. *Magasin de zoologie.* Paris, 1831+. Edited by F.E. Guérin-Méneville. Size: 1831-38, 9; 1839-49, 2. In 1849 it united with the *Revue zoologique* (0760) to form the *Revue et magasin de zoologie* (0764).
Scudder 1439. Bolton 2734.

0756. Zoological Society of London. (Founded 1826) See also 0757.
Proceedings. 1833+. Size: 1830-63, 4; 1864-73, 11; 1901-05, 11.
See Index. Scudder 484.

0757. Zoological Society of London. See also 0756.
Transactions. 1833 (1835)+. Size: 1835-65, 1; 1866-73, 2; 1901-05, 1.
Scudder 484.

0758. *Annales des sciences naturelles: Zoologie.* Paris, 1834+. (For

preceding series see 0602) Size: 1834-43, 7; 1844-53, 7; 1854-63, 5; 1864-73, 2; 1901-05, 1.
Scudder 1308. Bolton 377a & p. 629.

0759. *Archiv für Naturgeschichte*. Berlin, 1835+. Edited by A.F.A. Wiegmann. Size: 1835-63, 3; 1864-73, 2; 1901-05, 4. The latter half of each volume contained annual reviews of several aspects of zoology (and some botany). Some of these were published separately, notably 0768 and 0808, and the botanical items 0690 and 0694.
See Index. Scudder 2277, 2286-2289. Bolton 594, 876-890.

0760. *Revue zoologique*. Paris, 1838+. Published by the Société Cuviérienne. Edited by F.E. Guérin-Méneville. Size: 1838-48, 5. In 1849 it united with the *Magasin de zoologie* (0755) to form the *Revue et magasin de zoologie* (0764).
Scudder 1570. Bolton 1123.

0761. *The zoologist*. London, 1843+. Edited by E. Newman. Size: 1843-63, 4; 1864-73, 7; 1901-05. 11.
Scudder 485. Bolton 4946.

0762. State Cabinet (*from 1871* New York State Museum) of Natural History. Albany. See also 0786.
Annual report. 1847+. Size: 1901-05, 1.
Scudder 3950, 3946.

0763. *Zeitschrift für wissenschaftliche Zoologie*. Leipzig, 1848+. Edited by C.T. von Siebold and A. Kölliker. Size: 1849-63, 3; 1864-73, 3; 1901-05. 6.
See Index. Scudder 3046. Bolton 4891 & p. 998.

0764. *Revue et magasin de zoologie pure et appliquée*. Paris, 1849+. Formed by amalgamation of 0755 and 0760. Edited by F.E. Guérin-Méneville. Size: 1849-63, 7; 1864-73, 1. Ceased 1879.
Scudder 1526. Bolton 2734c & p. 824.

0765. Naturhistorisches Landes-Museum in Kärnten. Klagenfurt. (Publication record dates from 1811) See also 0793.
Jahrbuch. 1852+. Size: 1852-63, 1. Cont'd past 1900.
Scudder 3436. Bolton 2257.

0766. Royal Physical Society of Edinburgh. ("For the promotion of zoology and other branches of natural history") (Founded 1771)
Proceedings. 1854 (1858)+. Size: 1858-63, 1; 1864-73, 1; 1901-05, 1.
Scudder 122 (see also Errata).

0767. Linnean Society of London. See also 0595.
Journal (of the proceedings). Zoology. 1855 (1857)+. Size: 1857-63, 1; 1864-73, 2; 1901-05, 1.
Scudder 333.

0768. *Bericht über die wissenschaftlichen Leistungen in der Naturgeschichte der niederen Tiere*. Berlin, 1856 (1857)+. By R. Leuckart. Separate publication of one of the annual reviews included in the *Archiv für Naturgeschichte* (0759). Cont'd past 1900.
Scudder 2289. Bolton 890 & p. 661.

0769. *Der zoologische Garten; Zeitschrift für Beobachtung, Pflege und*

Zucht der Thiere. Frankfurt a.M., 1859 (1860)+. Published by the Zoologische Gesellschaft in Frankfurt a.M. (founded 1856). Size: 1860-73, 4; 1901-05, 6.
Scudder 2613. Bolton 4941.

0770. Museum of Comparative Zoology at Harvard College. Cambridge, Mass.
Bulletin. 1863+. Size: 1863-73, 1; 1901-05, 3.
Scudder 4021. Bolton 1157.

0771. *The record of zoological literature*. London, 1864-69 (1865-70). Edited by A.C.L.G. Günther (Keeper of Zoology at the British Museum of Natural History). Cont'd from 1870 (1871) as *The zoological record*. Published from 1871 by the Zoological Record Association, then from 1887 by the Zoological Society of London.
See Index. Scudder 429. Bolton 3876. Documentation: Neave, 1950.

0772. *Archives de zoologie expérimentale et générale: Histoire naturelle, morphologie, histologie, évolution des animaux*. Paris, 1872+. Edited by H. de Lacaze-Duthiers. Size: 1872-73, 2; 1901-05, 5.
Scudder 1335. Bolton 634 & p. 645.

0773. Nederlandsche Dierkundige Vereeniging. [Dutch Zoological Association]
Tijdschrift. Rotterdam, 1872 (1874)+. Size: 1901-05, 1.
Scudder 903. Bolton 4559, 8205.

0774. Société Zoologique de France. Paris. See also 0787.
Bulletin. 1876+. Size: 1901-05, 2.
Scudder 1606.

0775. Tromsø Museum.
Aarshefter. 1878+. Size: 1901-05, 1.

0776. *Zoologischer Anzeiger*. Leipzig, 1878+. Edited by J.V. Carus until his death in 1903. From 1890 published by the Deutsche Zoologische Gesellschaft (founded 1890; see 0794). Size: 1901-05, 11. Its *Litteratur* supplement, begun in 1891, became in 1896 a separate periodical, *Bibliographia zoologica*; see 0802.
Bolton 4943.

0777. *Le naturaliste: Revue illustrée des sciences naturelles*. Paris, 1879+. Edited by E. Deyrolle. Size: 1901-05, 6. Ceased 1910.
Bolton 7194.

0778. Rijksmuseum van Natuurlijke Historie te Leiden. (Muséum d'Histoire Naturelle des Pays-Bas) (Publication record dates from 1862)
Notes from the Leyden museum. 1879+. Size: 1901-05, 3.
Bolton 7273.

0779. American Museum of Natural History. New York. (Publication record dates from 1869)
Bulletin. 1881+. Size: 1901-05, 4.

0780. United States National Museum. (Smithsonian Institution) Washington. (Publication record dates from 1875)
Report, later *Annual report*. 1884+. Size: 1901-05, 2.

0781. *Recueil zoologique suisse.* Genève, 1884+. Edited by H. Fol. Cont'd from 1893 as *Revue suisse de zoologie et annales du Musée d'Histoire Naturelle de Genève.* Size: 1901-05, 2.
Bolton 7781.

0782. *Echange: Revue linnéenne.* Moulins, etc., 1885+. Size: 1901-05, 2 (all zoology).
Bolton 6037.

0783. K.-K. Naturhistorisches Hofmuseum. Wien. *Formerly* K.-K. Hof-Naturalien-Cabinet. *Formerly* Wiener Museum der Naturgeschichte. (Publication record dates from 1835)
Annalen. 1885 (1886)+. Size: 1901-05, 2.
Bolton 5228.

0784. R. Università di Torino. Musei di Zoologia e di Anatomia Comparata.
Bollettino. 1886+. Size: 1901-05, 2.
Bolton 5612.

0785. (a) *Zoologische Jahrbücher.* Jena, 1886+. Edited by J.W. Spengel. Cont'd from 1888 in two sections, as follows.
(b) ——— *Abteilung für Anatomie und Ontogenie der Tiere.* Size: 1901-05, 5.
(c) ——— *Abteilung für Systematik, Geographie und Biologie der Tiere.* Size: 1901-05, 3.
Bolton 8471.

0786. New York State Museum. Albany. See also 0762.
Bulletin. 1887+. Size: 1901-05, 3.

0787. Société Zoologique de France. Paris. See also 0774.
Mémoires. 1888+. Size: 1901-05, 1. Ceased 1932.

0788. *Bibliotheca zoologica: Original-Abhandlungen aus dem Gesammtgebiet der Zoologie.* Cassel, 1888+. Edited by R. Leuckart and C. Chun. Cont'd from 1897 as *Zoologica.* Size: 1901-05, 1.
Bolton 5542.

0789. International Zoological Congress. [Also in French, German, and Italian] (The first meeting was held at Paris in 1889)
Proceedings. 1889+.

0790. *North American fauna.* Washington, 1889+. Published by the Division of Biological Survey, U.S. Department of Agriculture. Size: 1901-05, 1.

0791. Australian Museum. Sydney. (Publication record dates from 1837)
Records. 1890 (1891)+. Size: 1901-05, 3.
Bolton 7646.

0792. *Monitore zoologico italiano.* Siena, 1890+. Published by the Unione Zoologica Italiana. Size: 1901-05, 5.
Bolton 7146.

0793. Naturhistorisches Landes-Museum in Kärnten. Klagenfurt. See also 0765.
Carinthia II: Mitteilungen.... 1891+. Size: 1901-05, 2.
Cf. Bolton 1211.
Carinthia I, begun at the same time, was a periodical for local history.

0794. Deutsche Zoologische Gesellschaft. (Founded 1890 in Frankfurt a.M. A *Wandergesellschaft* which held annual meetings in different German cities, beginning with Leipzig in 1891)
Verhandlungen. 1891+. Size: 1901-05, 2.

0795. R. Università di Genova. Museo di Zoologia ed Anatomia Comparata.
Bollettino. 1892 (1894)+. Size: 1901-05, 2.

0796. University of Pennsylvania. Zoological Laboratory.
Contributions. 1893+. Size: 1901-05, 2.

0797. *Novitates zoologicae: A journal of zoology*. London, 1894+. Published by the Tring Zoological Museum. Size: 1901-05, 4.
Bolton 7280.

0798. *Zoologisches Zentralblatt*. Leipzig, 1894+. Edited by A. Schuberg. Size: 1901-05, 1.
Bolton 8472.

0799. University of Nebraska. Zoological Laboratory.
*Studies (*sometimes *Contributions)*. Lincoln, 1894+. Size: 1901-05, 1.

0800. Muséum National d'Histoire Naturelle. Paris. See also 0753.
Bulletin. 1895+. Size: 1901-05, 10.
Bolton 5753.

0801. Imp. Akademiya Nauk. Zoologicheskii Muzei. (Académie Impériale des Sciences. Musée Zoologique) St. Petersburg.
Ezhegodnik (Annuaire). 1896+. Size: 1901-05, 3.

0802. *Bibliographia zoologica*. Leipzig, later Zürich, 1896+. Edited by J.V. Carus. Originated as a supplement to *Zoologischer Anzeiger*; see 0776. Later published by the Concilium Bibliographicum, Zürich.

0803. *Annotationes zoologicae Japonenses*. Tokyo, 1897+. Published (in European languages) by the Societas Zoologica Tokyonensis. Size: 1901-05, 1.

0804. South African Museum. Cape Town. (Publication record dates from 1855)
Annals. 1898 (1899)+. Size: 1901-05, 1.

(b) Entomology

0805. *Entomological magazine*. London, 1832 (1833)+. Edited by E. Newman. Cont'd from 1840 as *The entomologist* which from 1843 to 1863 was absorbed in *The zoologist* (0761; also edited by Newman) but from 1864 continued separately. Size: 1864-73, 2; 1901-05, 3.
Scudder 269, 272. Bolton 1582.

0806. Société Entomologique de France. Paris. See also 0834.
Annales. 1832+. Size: 1832-42, 5; 1843-52, 3; 1853-63, 9; 1864-73, 10; 1901-05, 2.
Scudder 1574.

0807. (Royal) Entomological Society of London. (Founded 1833)
Transactions. 1834 (1836)+. Size: 1836-49, 2; 1850-61, 2; 1862-73, 2; 1901-05, 5.
Scudder 271.

0808. *Bericht über die wissenschaftlichen Leistungen im Gebiete der Entomologie*. Berlin, 1838 (1840)+. Separate publication of one of the annual reviews included in the *Archiv für Naturgeschichte* (0759). Ceased 1913.
Scudder 2287. Bolton 882.

0809. Entomologischer Verein zu Stettin. (Founded 1837)
Jahresbericht. 1839. Cont'd from 1840 as *Entomologische Zeitung*. Size: 1840-63, 5; 1864-73, 3; 1901-05, 2.
Scudder 3225. Bolton 1587.

0810. *Zeitschrift für die Entomologie*. Leipzig, 1839+. Edited by E.F. Germer. Ceased in 1841 but revived in 1846 as *Linnaea entomologica*, published by the Entomologischer Verein zu Stettin (0809). Size: 1846-63, 1. Ceased 1866.
Scudder 2318. Bolton 4901.

0811. Nederlandsche Entomologische Vereeniging. (Publication record dates from 1845)
Handelingen. Leiden, 1854-57. Cont'd from 1857 as *Tijdschrift voor entomologie*. Size: 1858-63, 7; 1864-73, 1; 1901-05, 1.
Scudder 877, 833. Bolton 2030.

0812. Entomologischer Verein. Berlin. (Founded 1856)
Berliner entomologische Zeitschrift. 1857+. (Not to be confused with item 0823) Size: 1857-63, 8; 1864-73, 4; 1901-05, 2.
Scudder 2317. Bolton 896.

0813. Société Entomologique Belge. Bruxelles.
Annales. 1857+. Size: 1857-73, 2; 1901-05, 3.
Scudder 961.

0814. Societas Entomologica Rossica. (Russkoe Entomologicheskoe Obshchestvo) St. Petersburg.
Horae (Trudui). 1861+. Size: 1861-73, 2; 1901-05, 4.
Scudder 3747, 3740.

0815. Schweizerische Entomologische Gesellschaft.
Mittheilungen. Schaffhausen, 1862 (1865)+. Size: 1865-68, 3. Cont'd past 1900.
Scudder 2180.

0816. *The entomologist's monthly magazine*. London, 1864+. Edited by H.G. Knaggs et al. Size: 1864-74, 4; 1901-05, 6.
Scudder 274. Bolton 1593.

0817. *L'abeille: Mémoires d'entomologie*. Paris, 1864+. Edited by S.A. de Marseul. Size: 1864-73, 1; 1901-05, 1.
Scudder 1272. Bolton 6.

0818. Entomological Society of Philadelphia. (Founded 1861?) *From 1868* American Entomological Society.
Transactions. 1867 (1868)+. Size: 1867-73, 1; 1901-05, 1.
Scudder 4236.

0819. *The Canadian entomologist*. Toronto, 1868+. Edited by C.J.S. Bethune et al. From ca. 1875 published by the Entomological Society of Canada. Size: 1901-05, 2.
Scudder 4384. Bolton 1198.

0820. Società Entomologica Italiana. Firenze.
Bullettino. 1869+. Size: 1901-05, 1.
Scudder 1849.

0821. *Psyche: Journal of entomology*. Cambridge, Mass., 1874+. Published by the Cambridge Entomological Club. Size: 1901-05, 2.
Scudder 4019. Bolton 3808.

0822. *Entomologisk tidskrift*. Stockholm, 1880+. Published by the Entomologiska Förening. Size: 1901-05, 2.
Bolton 1591.

0823. Deutsche Entomologische Gesellschaft. (Founded in 1881 as an offshoot from the Entomologischer Verein in Berlin--0812)
Deutsche entomologische Zeitschrift. Berlin, 1881+. Size: 1901-05, 4.
Bolton 896a.

0824. United States. Department of Agriculture. Division of Entomology. Washington.
Bulletin. 1883+. Size: 1901-05, 3.

0825. Entomological Society of Washington.
Proceedings. 1884+. Size: 1901-05, 2.

0826. Entomologischer Verein 'Iris' zu Dresden. (Founded 1862)
Correspondenz-Blatt. 1884/88. Cont'd from 1889 as *Deutsche entomologische Zeitschrift: Lepidopterologische Heft*, published in association with the Deutsche Entomologische Gesellschaft (0823). Size: 1901-05, 2.
Bolton 5897.

0827. *Die Insectenbörse: Internationales Organ der Entomologie*. Leipzig, 1884+. Size: 1901-05, 4.
Bolton 6616.

0828. *Societas entomologica*. Zürich, 1886+. Published by the Internationaler Entomologenverein. Size: 1901-05, 3.
Bolton 8023.

0829. *Entomologische Zeitschrift*. Frankfurt a.M., 1887+. "Centralorgan des Internationalen Entomologischen Vereins ... und des Verbandes Deutschsprachlicher Entomologen-Vereine" Size: 1901-05, 5.
Bolton 6116.

0830. *The entomologist's record and journal of variation*. London, 1890 (1891)+. Edited by J.W. Tutt. Size: 1901-05, 2.
Bolton 6120.

0831. *Entomological news*. Philadelphia, 1890+. Published by the Entomological Section of the Academy of Natural Sciences of Philadelphia. Size: 1901-05, 6.
Bolton 6115.

0832. *Entomologisches Jahrbuch*. Leipzig, 1892+. Size: 1901-05, 2. Ceased 1937.
Bolton 6117.

0833. New York Entomological Society.
Journal. 1893+. Size: 1901-05, 3.

0834. Société Entomologique de France. Paris. See also 0806.
Bulletin. 1896+. Size: 1901-05, 7. Before 1896 it formed part of the *Annales* (0806).

0835. *Illustrierte Wochenschrift für Entomologie.* Neudamm, 1896+. Published by the Berliner Entomologische Gesellschaft. From 1898 *Illustrierte Zeitschrift für Entomologie.* Published by the Allgemeine Entomologische Gesellschaft. From 1901 *Allgemeine Zeitschrift für Entomologie.* Size: 1901-05, 6.
Bolton 6524.

(c) Ornithology

0836. *Naumannia: Archiv für die Ornithologie.* Stuttgart, 1849+. Published by the Deutscher Ornithologen-Verein (founded 1845). Size: 1849-58, 3. In 1859 it was incorporated in the *Journal für Ornithologie* (0837).
Scudder 3246. Bolton 3286.

0837. *Journal für Ornithologie.* Cassel, 1853+. Published by the Deutsche Ornithologen-Gesellschaft (*formerly* Verein). Size: 1853-73, 4; 1901-05, 3. Supplement: see 0850.
Scudder 2467, 2468. Bolton 2483.

0838. *The ibis: A magazine of general ornithology.* London, 1859+. Edited by P.L. Sclater. Size: 1859-63, 5; 1864-72, 3; 1901-05, 5.
Scudder 302. Bolton 2102 & p. 771.

0839. *Die gefiederte Welt.* Magdeburg, 1872+. Edited by K. Russ. Size: 1901-05, 3.
Scudder 2319. Bolton 1856.

0840. Ornithologischer Verein in Stettin. (Founded 1873) *From 1883* Verband der Ornithologischen Vereine Pommerns und Mecklenburgs.
Jahresbericht. 1874+. Cont'd from 1877 as *Zeitschrift.* Cont'd from 1883 as *Zeitschrift für Ornithologie und praktische Geflügelzucht.* Size: 1901-05, 2.
Bolton 8400.

0841. Nuttall Ornithological Club. Cambridge, Mass. (Named after Thomas Nuttall (1786-1859), an English-American naturalist and ornithologist)
Bulletin. 1876+. Cont'd from 1884 as *The Auk: Quarterly journal of ornithology.* Size: 1901-05, 8.
Scudder 4022. Bolton 5755.

0842. Sächsisch-Thüringischer Verein für Vogelkunde und Vogelschutz zu Halle a.d. Salle. *Later* Deutscher Verein zum Schutze der Vogelwelt.
Monatsschrift. 1876+. From 1890 *Ornithologische Monatsschrift.* Size: 1901-05, 6.
Bolton 7379.

0843. *Schweizerische Blätter für Ornithologie.* Zug, 1877+. Published by the Schweizerischer Ornithologischer Verein. Size: 1901-05, 1.
Bolton 4236.

0844. International Ornithological Congress. [Also in French and German] (The first meeting was held at Vienna in 1884)
Proceedings. 1884+.

0845. *Ornis: Internationale Zeitschrift für die gesammte Ornithologie.* Wien, 1885+. "Organ des Permanenten Internationalen Ornithologischen Comité's" From 1898 published from Paris as *Ornis: Bulletin du Comité ornithologique international.* Size: 1901-05, 1. Ceased 1910.
Bolton 7376.

0846. *Ornithologisches Jahrbuch: Organ für das palaearktische Faunengebiet.* Hallein, 1890+. Size: 1901-05, 1. Ceased 1918.
Bolton 7380.

0847. *Zeitschrift für Oologie (*later *und Ornithologie).* Berlin, 1891+. Size: 1901-05, 1. Ceased 1924.
Bolton 8436.

0848. British Ornithologists' Club. (Founded 1892)
Bulletin. London, 1892/93+. Size: 1901-05, 6.

0849. The Wilson Ornithological Chapter of the Agassiz Association. (Named after Alexander Wilson (1766-1813), an outstanding American ornithologist)
The Wilson quarterly. Oberlin, Ohio, 1892. Cont'd from 1893 as *Journal.* Cont'd from 1894 as *The Wilson bulletin.* Size: 1901-05, 2.

0850. *Ornithologische Monatsberichte.* Berlin, 1893+. A supplement to the *Journal für Ornithologie* (0837). Size: 1901-05, 8.
Bolton 7378.

0851. *The avicultural magazine.* Brighton, 1894+. Published by the Avicultural Society. Size: 1901-05, 3.

0852. *Avicula: Giornale ornitologico italiano.* Siena, 1897+. Size: 1901-05, 1. Ceased 1910.
Bolton 8526.

(d) Malacology and Conchology

0853. *Zeitschrift für Malakozoologie.* Hannover, 1844 (1845)+. Edited by K.T. Menke and L. Pfeiffer. Size: 1844-53, 2. Cont'd from 1854 as *Malakozoologische Blätter.* Size: 1854-63, 1; 1864-73, 2. Ceased 1891.
Scudder 2791, 2469. Bolton 4861.

0854. *Journal de conchyliologie.* Paris, 1850+. Edited by S. Petit de la Saussaye. Size: 1850-63, 7; 1864-73, 2; 1901-05, 2.
Scudder 1405. Bolton 2368.

0855. Société Malacologique (*from 1903* Société Royale Zoologique et Malacologique) de Belgique. Bruxelles.
Annales. 1863 (1865)+. Size: 1863-73, 1; 1901-05, 1.
Scudder 963.

0856. Deutsche Malakozoologische Gesellschaft. Frankfurt a.M. (Founded 1868)
Nachrichtsblatt. 1868 (1869)+. Size: 1901-05, 2.
Scudder 2585.

0857. *Quarterly journal (*from 1879 *Journal) of conchology.* Manchester, etc., 1874+. Published by the Conchological Society of Great

Britain and Ireland. Size: 1901-05, 2.
Scudder 155. Bolton 3824.

0858. *The conchologists' exchange*. Philadelphia, 1886+. Cont'd from 1889 as *The nautilus*. Size: 1901-05, 6.
Bolton 5883.

0859. *The conchologist*. London, 1891+. Cont'd from 1894 as *The journal of malacology*. Size: 1901-05, 2. Ceased 1905.
Bolton 5882.

0860. The Malacological Society of London. (Founded 1893)
Proceedings. 1893+. Size: 1901-05, 4.

(e) Ichthyology

0861. *Mittheilungen über Fischereiwesen*. München, 1876+. Published by the Bayerischer und Deutscher Fischereiverein. Cont'd from 1879 as *Bayerische Fischerei-Zeitung*. Cont'd from 1886 as *Allgemeine Fischereizeitung*. Size: 1901-05, 1.
Scudder 3111. Bolton 3086.

0862. United States. Fish Commission. Washington.
Bulletin. 1881 (1882)+. Size: 1901-05, 3.

0863. *Okólnik rybacki*. [*Journal of ichthyology*] Krakowie, 1895+. Published by the Krajowego Towarzystwa Rybackiego w Krakowie [Regional Society for Ichthyology in Cracow] Size: 1901-05, 1.

(f) Marine Zoology

See also Oceanography.

0864. Zoologische Station zu Neapel.
Mittheilungen ... zugleich ein Repertorium für Mittelmeerkunde. Leipzig, etc., 1878+. Size: 1901-05, 2.
Bolton 3081.
The Zoological Station at Naples was established in 1874 by F.A. Dohrn. "It was not only the first laboratory set up specifically for marine studies but also the first institute formally organized for the sole pursuit of research and the prototype of those that followed."--DSB (article on Dohrn, in Vol. XV)

0865. Marine Biological Association of the United Kingdom. Plymouth. (Founded 1884)
Journal. London, later Plymouth, 1887+. Size: 1901-05, 1.

0866. Biologische Station zu Plön. [Plön is on a lake near the Baltic, between Kiel and Lübeck]
Forschungsberichte. Berlin, 1893+. Size: 1901-05, 3. Cont'd from 1905 as *Archiv für Hydrobiologie und Planktonkunde*.

0867. Société Scientifique d'Arcachon. [Arcachon is on the Atlantic coast, near Bordeaux]
Travaux des laboratoires de la Société scientifique et Station zoologique d'Arcachon. Bordeaux, 1895+. Size: 1901-05, 1.

0868. Marine Biological Laboratory. Woods Hole, Mass. (Founded 1888)
Biological bulletin. Boston, 1898+. Size: 1901-05, 4.

EXPERIMENTAL BIOLOGY

Covering anatomy, physiology, and other aspects of experimental biology. (Biochemistry is included under Chemistry)

Including some periodicals for pathological anatomy and physiology which contained substantial numbers of papers bearing on normal anatomy and physiology. Medical periodicals are excluded.

0869. (a) *Archiv für die Physiologie*. Halle, 1795 (1796)+. Edited by J.C. Reil. Size: 1796-1815, 1. Cont'd as (b).
(b) *Deutsches Archiv für die Physiologie*. Ib., 1815+. Edited by J.F. Meckel. Size: 1815-23, 1. Cont'd from 1826 as *Archiv für Anatomie und Physiologie*. Leipzig, 1826+. Edited by Meckel. Size: 1826-32, 3. Ceased 1832. Succeeded by 0872.
See Index. Scudder 2705, 2710, 2888. Bolton 611. Documentation: Geus, 1971.

..... *Notizen aus dem Gebiete der Natur- und Heilkunde*. 1821+. Main entry 076.

0870. *Journal de physiologie expérimentale* (from Vol. 2 *et pathologique*). Paris, 1821+. Edited by F. Magendie. Size: 1821-31, 1. Ceased 1831.
See Index. Scudder 1419. Bolton 2387.

0871. Société Anatomique. Paris.
*Bulletins (*later *et mémoires)*. 1826+. Size: 1826-61, 1; 1864-73, 1; 1901-05, 5.
See Index. Scudder 1554.

0872. (a) *Archiv für Anatomie, Physiologie und wissenschaftliche Medicin*. Berlin, 1834+. (Preceded by 0869) Edited by J. Müller until his death in 1858, thereafter until 1876 by C.B. Reichert and E. Du Bois-Reymond. Size: 1834-58, 6; 1859-73, 6. From 1877 divided into two sections, as follows.
(b) ——— *Archiv für Anatomie und Entwickelungsgeschichte. Anatomische Abtheilung des "Archives für Anatomie und Physiologie."* Leipzig, 1877+. Edited by W. His and W. Braune. Size: 1901-05, 4.
(c) ——— *Archiv für Physiologie. Physiologische Abtheilung des "Archives für Anatomie und Physiologie."* Ib., 1877+. Edited by E. Du Bois-Reymond. Size: 1901-05, 8.
See Index. Scudder 2273. Bolton 611.

0873. *(Canstatt's) Jahresberichte über die Fortschritte in der Biologie*. Erlangen, 1843 (1844)+. Edited by C. Canstatt and J.G. Eisenmann. Cont'd from 1851 (1852) as *Canstatt's Jahresbericht über die Leistungen in den physiologischen Wissenschaften*. Ceased 1873.
Scudder 2564. Bolton 1209.
This periodical formed the first section of the *Jahresbericht über die Fortschritte der gesammten Medicin in allen Ländern*. Erlangen, 1841-72. Also edited by Canstatt.

0874. *Archiv für pathologische Anatomie und Physiologie*. Berlin, 1847+.

Edited by R. Virchow until his death in 1902. Size: 1847-63, 3; 1864-73, 5; 1901-05, 11.
See Index. Bolton 596 & p. 643.

0875. Utrechtsche Hoogeschool [Utrecht University]. See also 0415.
Onderzoekingen gedaan in het physiologisch laboratorium. Rotterdam, 1848+. Size: 1854-56, 1; 1872-73, 1; 1901-05, 2.
Scudder 907. Bolton 7350.

0876. Société de Biologie. Paris. (Founded 1848)
Comptes rendus (later *hebdomadaires*) *des séances et mémoires.* 1849 (1850)+. Size: 1849-63, 5; 1864-73, 5; 1901-05, 16.
See Index. Scudder 1564.

0877. (a) *Bericht über die Fortschritte der Anatomie und Physiologie.* Leipzig, etc., 1856 (1857)+. Edited by J. Henle and G. Meissner. (Issued until 1868 as a supplement to the *Zeitschrift für rationelle Medicin*). Cont'd as (b).
(b) *Jahresberichte über die Fortschritte der Anatomie und Physiologie.* Leipzig, 1872 (1873)+. Edited by F. Hofmann and G. Schwalbe. From 1892 (1893) divided into two sections, as follows.
(c) ——— *Jahresbericht über die Fortschritte der Anatomie und Entwicklungsgeschichte.* Jena, 1892 (1893)+. Edited by G. Schwalbe. Ceased 1914.
(d) ——— *Jahresbericht über die Fortschritte der Physiologie.* Bonn, etc., 1892 (1894)+. Edited by L. Hermann. Ceased 1919.
Scudder 2905, 2944. Bolton 893, 4882 & p. 997.

0878. *Untersuchungen zur Naturlehre des Menschen und der Thiere.* Frankfurt a.M., 1857+. Edited by J. Moleschott. Size: 1857-63, 4; 1864-73, 1; 1901-05, 1.
Scudder 2609. Bolton 4638.

0879. *Lo sperimentale.* Firenze, 1858+. Issued by the Accademia Medico-Fisica Fiorentina. Size: 1864-73, 1; 1901-05, 3.

0880. *Beiträge zur Anatomie und Physiologie.* Giessen, 1858+. Edited by C. Eckhard. Size: 1865-74, 1. Ceased 1888.
Scudder 2636. Bolton 801.

0881. *Journal de la physiologie de l'homme et des animaux.* Paris, 1858+. Edited by E. Brown-Séquard. Size: 1858-63, 2. Ceased in 1863 but resumed in 1868 as *Archives de physiologie normale et pathologique.* Size: 1868-73, 3. Cont'd from 1899 as *Journal de physiologie et de pathologie générale.* Size: 1901-05, 2.
Scudder 1420, 1333. Bolton 2403 & p. 796/7.

0882. *Journal de l'anatomie et de la physiologie normales et pathologiques de l'homme et des animaux.* Paris, 1864+. Edited by C. Robin. Size: 1864-73, 3; 1901-05, 2. Ceased 1919.
Scudder 1400. Bolton 2396.

0883. *Archiv für mikroskopische Anatomie* (from 1895 *und Entwicklungsgeschichte*). Bonn, 1865+. Edited by M. Schultze until his death in 1874, later by O. Hertwig et al. Size: 1865-73, 4; 1901-05, 11.
See Index. Scudder 2406. Bolton 593, 5396.

0884. *Zeitschrift für Biologie.* München, 1865+. Edited by M. Petten-

kofer, C. Voit, et al. Size: 1865-73, 2; 1901-05, 4.
Scudder 3127. Bolton 4844 & p. 992.

0885. Leipzig. Universität. Physiologische Anstalt.
Arbeiten. 1866 (1867)+. Edited by C. Ludwig. Size: 1867-74, 1. Ceased 1876 (1877).
Scudder 2989. Bolton 534.
Most of the papers from Ludwig's famous institute appeared in other periodicals, especially the *Archiv für Physiologie* (0872c)

0886. Würzburg. Universität. Physiologisches Laboratorium.
Untersuchungen. Leipzig, 1867-69. Cont'd from 1872 as *Arbeiten*. Edited by A. Fick. Size not determined. Ceased 1878.
Scudder 2991, 3331.

0887. *Journal of anatomy and physiology*. London, 1867+. Edited by W. Turner, G.M. Humphrey, et al. Size: 1867-73, 3; 1901-05, 5.
See Index. Scudder 318. Bolton 2500 & p. 802. Documentation: Blake, 1968; Walls, 1966.
Despite its title it was from the 1870s a journal of anatomy only. When the Anatomical Society (founded 1889) assumed responsibility for it in 1916 the words "*and physiology*" were dropped.

0888. *Jahresbericht über die Leistungen und Fortschritte in der Anatomie und Physiologie*. Berlin, 1868 (1869)+. Edited by R. Virchow and A. Hirsch. Cont'd past 1900.
Scudder 2335. Bolton 2301.

0889. *Archiv für die gesammte Physiologie des Menschen und der Thiere*. Bonn, 1868+. Edited by E.F.W. Pflüger until his death in 1910. Size: 1868-73, 4; 1901-05, 12.
See Index. Scudder 2405. Bolton 604.

0890. Università di Roma. Laboratorio di Anatomia Normale.
Richerche fatte nel laboratorio. 1872+. Size: 1901-05, 1.

0891. *Archiv für experimentelle Physiologie und Pharmakologie*. Leipzig, 1873+. Edited by E. Klebs, B. Naunyn, and O. Schmiedeberg. Size: 1901-05, 5.

0892. *Morphologisches Jahrbuch: Eine Zeitschrift für Anatomie und Entwickelungsgeschichte*. Leipzig, 1875 (1876)+. Edited by C. Gegenbauer. Size: 1901-05, 3.
Scudder 2970. Bolton 3180 & p. 851.

0893. *Journal of physiology*. London/Cambridge, 1878 (1879)+. Edited by M. Foster until 1894. Published from 1884 by the Physiological Society (founded 1876). Size: 1901-05, 7.
Bolton 2519.

0894. *Brain: A journal of neurology*. London, 1878 (1879)+. Edited by D. Ferrier et al. Published from 1888 by the Neurological Society of London. Size: 1901-05, 3.
Bolton 1057.

0895. Biological Society of Washington.
Proceedings. 1880/82 (1883)+. Size: 1901-05, 5.

0896. Société d'Anatomie et de Physiologie Normales et Pathologiques de Bordeaux.
Bulletin. 1880+. Size: 1901-05, 1. Ceased 1913.

0897. *Biologisches Centralblatt*. Erlangen, 1881+. Edited by J. Rosenthal. Size: 1901-05, 10. Besides original papers it included a supplement containing abstracts.
Bolton 974.

0898. *Archives italiennes de biologie. (Archivio italiano di biologia)* Turin, 1882+. Edited by A. Mosso. Size: 1901-05, 11.
See Index. Bolton 625.

0899. *Neurologisches Zentralblatt*. Leipzig, 1882+. Size: 1901-05, 3.

0900. (a) *Internationale Monatsschrift für Anatomie und Histologie (from 1888 für Anatomie und Physiologie)*. Berlin/Paris/London, 1884+. Edited by E.A. Schäfer, L. Testut, and W. Krause. Size: 1901-05, 3. Ceased 1916 (1918).
(b) ——— [English edition] *Monthly international journal of anatomy and histology (from 1887 of anatomy and physiology)*. 1884+. Ceased 1899.
Bolton 6624.

0901. *Beiträge zur pathologischen Anatomie und Physiologie*. Jena, 1884+. Edited by E. Ziegler. Cont'd from 1888 as *Beiträge zur pathologischen Anatomie und zur allgemeinen Pathologie*. Size: 1901-05, 7.
Bolton 5521.

0902. Gesellschaft für Morphologie und Physiologie in München. (Founded 1875)
Sitzungsberichte. 1885 (1886)+. Size: 1901-05, 1.

0903. *Anatomischer Anzeiger*. Jena, 1886+. Edited by K. Bardeleben. Published by the Anatomische Gesellschaft (founded 1886 as a *Wandergesellschaft*). Size: 1901-05, 13.
See Index. Bolton 5223.

0904. Liverpool Biological Society.
*Proceedings (*later *and transactions)*. 1886 (1887)+. Size: 1901-05, 2.

0905. *Centralblatt für Physiologie*. Leipzig, 1887+. Published by the Physiologische Gesellschaft zu Berlin (founded 1875). A bibliographical periodical. In 1921 it was incorporated in *Pflüger's Archiv* (0889).
Bolton 5821.

0906. *Journal of morphology*. Boston, Mass., 1887+. Edited by C.O. Whitman and E.P. Allis. Suspended 1901-08. Size not determined.
Bolton 6784.

0907. *Skandinavisches Archiv für Physiologie*. Leipzig, 1889+. Edited by F. Holmgren. Size: 1901-05, 3.
Bolton 8013.

0908. International Physiological Congress. [Also in French and German]
[*Proceedings*] Basel, etc., 1889+.

0909. *Centralblatt für allgemeine Pathologie und pathologische Anatomie*. Jena, 1890+. Published by the Deutsche Pathologische Gesellschaft. Size: 1901-05, 9.
Bolton 5817.

0910. *Journal of comparative neurology.* Philadelphia, 1891+. Edited by C.L. Herrick. Size: 1901-05, 2.
Bolton 6778.

0911. *Anatomische Hefte.* Wiesbaden, etc., 1891+. Size: 1901-05, 10.
Bolton 5222.

0912. *Morphologische Arbeiten.* Jena, 1891 (1892)+. Edited by G. Schwalbe. Cont'd from 1899 as *Zeitschrift für Morphologie und Anthropologie.* Size: 1901-05, 3.
Bolton 7156.

0913. *Arkhiv biologicheskikh nauk. (Archives des sciences biologiques)* St. Petersburg, 1892+. Published by the Imp. Institut Eksperimentalnoi Meditsinui. Size: 1901-05, 3.
Bolton 5413.

0914. *Bibliographie anatomique: Revue des travaux en langue française.* Paris, 1893+. Edited by A. Nicolas. Size: 1901-05, 3.

0915. *Bibliographia physiologica: Répertoire des travaux de physiologie de l'année.* Paris, later Bruxelles, etc., 1893 (1894)+. Published from 1897 (1898) by the Institutus Bibliographicus Bruxellensis, then from 1905 by the Concilium Bibliographicum.

0916. *Archiv für Entwicklungsmechanik der Organismen.* Berlin, 1894+. Edited by W. Roux. Size: 1901-05, 9.
See Index. Bolton 5393.

0917. *L'année biologique: Comptes rendus annuels des travaux de biologie générale.* Paris, 1895 (1897)+. Edited by Y. Delage.
Bolton 8486.

0918. *Bibliographia anatomica.* Zürich, 1896+. Published by the Concilium Bibliographicum.

0919. *Bibliographia biologica.* Zürich, 1896+. Published by the Concilium Bibliographicum.

0920. Sankt-Peterburgskaya Biologicheskaya Laboratoriya.
Izvestiya. 1896+. Size: 1901-05, 2.

0921. *Archives d'anatomie microscopique.* Paris, 1897+. Edited by E.G. Balbiani, L. Ranvier, and L.F. Henneguy. Size: 1901-05, 2.
Bolton 8496.

0922. *American journal of physiology.* Boston, 1898+. Published by the American Physiological Society (founded 1887). Size: 1901-05, 8.

0923. Association des Anatomistes. (Held annual meetings in different French cities)
Comptes rendus. 1899+. Issued as a supplement to *Bibliographie anatomique* (0914). Size: 1901-05, 3.

MICROBIOLOGY

Including some periodicals for "hygiene" and applied microbiology. Medical periodicals are excluded.

0924. *Beiträge zur Biologie der Pflanzen.* Breslau, 1870+. Edited by F. Cohn. Size: 1901-05, 1.
See Index. Scudder 2441. Bolton 805. Documentation: DSB--article on Cohn. ("In this journal appeared the founding papers of modern bacteriology.")

0925. *Archiv für Hygiene (*later *und Bakteriologie).* München, 1883+. Edited by M. Pettenkofer. Size: 1901-05, 4.
Bolton 5394. Documentation: DSB--article on Pettenkofer.

0926. *Jahresbericht über die Fortschritte und Leistungen auf dem Gebiete der Hygiene.* Braunschweig, 1883+.
Bolton 6696.

0927. *Jahresbericht über die Fortschritte in der Lehre von den pathogenen Mikroorganismen, umfassend Bakterien, Pilze, und Protozoën.* Braunschweig, 1885+. Edited by P. Baumgarten. Ceased 1911.
Bolton 6694.

0928. *Zeitschrift für Hygiene.* Leipzig, 1886+. Edited by R. Koch and C. Flügge. Cont'd from 1892 as *Zeitschrift für Hygiene und Infectionskrankheiten.* Size: 1901-05, 8.
Bolton 8422.

0929. Institut Pasteur. Paris. (Founded 1888)
Annales. 1887 (1888)+. Edited by E. Duclaux. Size: 1901-05, 4.
Bolton 5245.

0930. *Centralblatt für Bakteriologie und Parasitenkunde (*later *und Infektionskrankheiten).* Jena, 1887+. Edited by O. Uhlworm. Size: 1901-05, 15. Included abstracts as well as original papers.
Bolton 5818.

0931. *Annali d'igiene sperimentale.* Roma, 1891+. Size: 1901-05, 1.

0932. *Jahresbericht über die Fortschritte in der Lehre von den Gährungs-Organismen.* Braunschweig, 1891+. By A. Koch.
Bolton 6693.

0933. *Journal of pathology and bacteriology.* Edinburgh, etc., 1892 (1893)+. Published by the Pathological Society of Great Britain and Ireland. Size: 1901-05, 2.
Bolton 6786.

0934. *Saikingaku zasshi.* [*Journal of bacteriology*] Tokyo, 1895+. In Japanese. Size: 1901-05, 1.

0935. *Russky arkhiv patologii, klinicheskoi meditsinui i bakteriologii. (Archives russes de pathologie, de médecine clinique et de bactériologie).* St. Petersburg, 1896+. Size: 1901-05, 4.

0936. *Archives de parasitologie.* Paris, 1898+. Size: 1901-05, 3.

PART 2

SELECTION OF THE PERIODICALS AND THEIR RELATIVE SIZES

2.1 The Seventeenth and Eighteenth Centuries

As mentioned in the Introduction, the criterion adopted for the inclusion of periodicals in the Catalogue is the extent of their use as means of publication by the scientific community at the relevant time. To determine this use, counts can be made of the references recorded in an index of the periodical literature of science, and for the nineteenth century the Royal Society's *Catalogue of Scientific Papers* was used in this way, as is described in the following section. For the period 1665-1800 the only major periodical index for the scientific literature is J.D. Reuss' *Repertorium Commentationum a Societatibus Litterariis Editarum* (1801-21) which has the severe disadvantage that it does not cover non-institutional periodicals. In Reuss' time they were regarded as inferior to the publications of the academies and other learned societies, as indeed most of them were. As we now know, however, some were quite important.

In order to include the non-institutional periodicals a different approach was adopted for the period before 1800. A list was compiled of the most important scientists of the period for whom bibliographical data are available in J.C. Poggendorff's *Biographisch-literarisches Handwörterbuch* (1863), and from the references given therein a list was drawn up of the periodicals used by these scientists, with a count of the number of scientists using each of the periodicals.

The list of scientists was obtained in the first place by extracting from my *Historical Catalogue of Scientists and Scientific Books* the names of all those born between 1600 and 1780, i.e. those who were active in the period 1665-1800 -- a total of 3811 persons. Of these, 1120 (29%) are not listed in Poggendorff's *Handwörterbuch*; these were nearly all botanists and zoologists with a few anatomists, since Poggendorff's work is restricted to the non-biological sciences (it does include some biologists but not biological literature). Of the remaining 2691 persons who are included in Poggendorff not all published their work in periodicals. As can be seen from Table 1 the proportion who did so increases from 17% of those born in the decade 1600-09 to 85% of those born in 1770-79 (these figures relate to the number of persons for whom Poggendorff records periodical articles; he does not claim that his bibliographies are complete).

This leaves 1752 persons with periodical articles recorded in Poggendorff and these constitute our sample. They include virtually all the leading scientists of the period with the major exception that only a few biologists are represented. The lack of biologists (who

Table 1

Proportion of Scientists Publishing in Periodicals

Decade of Birth	Number of Scientists	Publishing in Periodicals No.	%
1600-09	42*	7	17
1610-19	43*	13	30
1620-29	74*	29	39
1630-39	69	29	42
1640-49	79	34	43
1650-59	74	35	47
1660-69	70	40	57
1670-79	104	55	53
1680-89	95	58	61
1690-99	118	74	63
1700-09	139	90	65
1710-19	192	129	67
1720-29	223	136	61
1730-39	236	172	73
1740-49	279	209	75
1750-59	264	199	75
1760-69	270	211	78
1770-79	273	232	85

*Not counting those who died before 1665.

in this period were mostly naturalists) is not such a serious defect in the sample as might at first appear because in this period we are concerned with periodicals covering all or most of the sciences, rather than specialized periodicals, and it can be assumed that the general science periodicals chiefly used by non-biologists were the same as those chiefly used by biologists.

With this assumption it is possible to make numerical estimates of the degree of use of the general science periodicals of the time based on the publication practices of the sample of 1752 scientists. Table 2 lists the forty-one most used periodicals beginning before 1790, arranged by country or region, with the number of scientists in the sample who published one or more articles (often many articles) in them before 1800. The periodicals began at different times, of course, and some of them ceased publication during the eighteenth century; in order to obtain a comparative measure the number of authors is divided by the number of years the periodical ran before 1800. The number so obtained is only an index based on the sampling procedure described above, but it enables the periodicals to be ranked according to their degree of use and distinguishes those to be included in the Catalogue from those to be excluded.

The number of periodicals beginning before 1790 which were excluded totalled 158. 51% of them were German, 18% Italian, 12% French, 3% British, and 16% various other nationalities. All of them contained one or more articles written by scientists in the sample used but not enough such articles to warrant their inclusion in the Catalogue, or alternatively they were too short-lived irrespective of the number of articles. They fall into two roughly equal categories -- general

Table 2

The Chief Periodicals Beginning Before 1790

	Period[a]	No. of Authors	Annual Average
France			
Journal des sçavans (01)	1665-	b	
Académie des Sciences *Histoire et mémoires* (05)	1699-1789[c]	184	2.02
Mémoires de Trevoux (06)	1701-1767	14	0.21
Académie de Bordeaux *Recueil ... prix* (09)	1715-1735?	8	0.38?
Académie des Sciences *Recueil ... prix* (010)	1720-1772	17	0.32
Académie des Sciences *Mém. sav. étr.* (018)	1750-1786	58	1.57
Rozier's Journal (030)	1771-(1799)	101	3.48
Académie de Toulouse *Histoire et mémoires* (036)	1782-1790	5	0.56
Britain			
Philosophical transactions (02)	1665-(1799)	334	2.47
Manchester Lit. & Phil. Soc. *Memoirs* (039)	1785-(1799)	8	0.53
Royal Irish Academy *Transactions* (043)	1787-(1799)	7	0.54
Royal Society of Edinburgh *Transactions* (044)	1788-(1799)	15	1.25
Germany			
Academia Naturae Curiosorum *Miscellanea* (03)	1670-(1799)	92	0.71
Acta eruditorum (06)	1682-1776	46	0.48
[Berlin Academy] (07 and 017)	1710-(1799)	84	0.93
Göttingische Zeitung (015)	1739-(1799)	9	0.15
[Göttingen Society] *Commentarii* (020)	1751-(1799)	17	0.35
[Erfurt Academy] *Acta* (024)	1757-(1799)	27	0.63
[Munich Academy] *Abhandlungen* (027)	1763-1797	23	0.66
[Mannheim Academy] *Hist. et comm.* (028)	1766-1794	9	0.31
Der Naturforscher (0593)[d]	1774-(1799)	(10)	(0.38)
Gesell. Naturf. Freunde,[d] Berlin *Beschäftigungen* (0594)[d]	1775-(1799)	(60)	(2.40)
Mag. Neueste Physik und Naturgeschichte (035)	1781-(1799)	17	0.89
Italy			
Giornale dei letterati (08)	1710-1740	9	0.29
[Bologna Institute] *Commentarii* (013)	1731-(1799)	32	0.46

[Turin Academy] (025 and 038)	1759-1773] 1784-(1799)]	36	1.16
[Siena Academy] *Atti* (026)	1761-(1799)	13	0.33
Scelta di opuscoli (032)	1775-(1799)	14	0.56
Soc. Ital. dei Quaranta *Memorie* (037)	1782-(1799)	34	1.89
[Padua Academy] *Saggi* (042)	1786-1794	5	0.56
The Low Countries			
[Harlem Society] *Verhandelingen* (023)	1754-(1799)	20	0.43
[Rotterdam Society] *Verhandelingen* (031)	1774-(1799)	6	0.23
[Brussels Academy] *Mémoires* (034)	1777-1788	8	0.67
Scandinavia			
[Uppsala Society] *Acta* (011)	1720-(1799)	24	0.30
[Stockholm Academy] *Handlingar* (014)	1739-(1799)	73	1.20
[Copenhagen Society] *Skrifter* (016)	1743-1799	18	0.32
Switzerland			
Acta Helvetica (019)	1751-1777	13	0.76
Eastern Europe			
[Prague Society] (033 and 040)	1775-1784] 1785-(1799)]	23	0.92
Russia			
[St. Petersburg Academy] *Commentarii* (012)	1726-(1799)	69	0.93
America			
American Philosophical Soc. *Transactions* (029)	1769-(1799)	14	0.45
American Academy of Arts & Sci. *Memoirs* (041)	1785-(1799)	3	0.20

Notes: (a) Where a periodical continued beyond 1800 (the usual case) only those authors who published papers in it before that year were counted. In these cases the end of the period is for present purposes taken to be 1799.

(b) Up to 1700 the *Journal des Sçavans* was used by sixteen of the authors in the sample, thereafter by very few.

(c) At the suppression of the Academy in August 1793 the last volume that had been published (earlier in 1793) was the one for 1789.

(d) A natural history periodical. Because of the lack of naturalists in the sample the figures for this periodical are much lower than they should be. (The references to it in Poggendorff deal with mineralogy, which then included geology and some chemistry.)

Table 3

Examples of Short-Lived Scientific Journals Excluded from the Catalogue

The editors mentioned are included in Poggendorff's *Handwörterbuch* and those with an asterisk are also in the *Dictionary of Scientific Biography*. Some of the journals contained articles by important scientists but there is always the possibility that such articles were reprinted from other sources.

Acta medica et philosophica Hafniensia. 5 vols. 1671-79. Ed. by T. Bartholin*.

Breslauische Sammlungen der Natur und Kunst. 38 parts. 1718-28. Ed. by J.C. Kundmann et al. Contained articles by seven of the scientists in the survey.

Miscellanea physico-medico-mathematica. 4 vols. 1731-34. Ed. by A.E. Büchner.

Recueil pour les astronomes. 3 vols. 1772-76. Ed. by Johann (III) Bernoulli*.

Journal für Liebhaber des Steinreichs und der Conchyliologie. 6 vols. 1773-80. Ed. by J.S. Schröter.

Mineralogischer Briefwechsel. 3 vols. 1779-84. Ed. by P.E. Klipstein.

Göttingisches Magazin für Wissenschaft und Literatur. 3 vols. 1780-82. Ed. by G.C. Lichtenberg* and J.G.A. Forster*. Contained articles by ten of the scientists in the survey as well as by the two editors.

Leipziger Magazin für Naturkunde, Mathematik und Oekonomie. 5 vols. 1781-85. Ed. by K.F. Hindenburg*, C.B. Funk, and N.G. Leske.

Neue Beiträge zur Natur- und Arzeneywissenschaft. 3 vols. 1782-86. Ed. by C.G. Selle. Contained articles by M.H. Klaproth and others.

Physikalische Arbeiten der einträchtigen Freunde. 2 vols. 1783-88. Ed. by I. von Born*. Contained articles by nine of the scientists in the survey as well as by von Born.

Physikalisches Tagebuch. 1784-87. Ed. by L. Hübner.

Magazin für Apotheker, Materialisten und Chemisten. 1785-87. Ed. by J.K.P. Elwert.

Leipziger Magazin für reine und angewandte Mathematik. 4 parts. 1786-88. Ed. by K.F. Hindenburg* and Johann (III) Bernoulli*. Contained articles by H.W.M. Olbers, J.H. Lambert, and several others in the survey.

Magazin für die Naturkunde Helvetiens. 4 vols. 1787-89. Ed. by J.G.A. Höpfner. Contained articles by J.R. Forster, A.G. Werner, M.H. Klaproth, and several others in the survey.

Biblioteca fisica d'Europa. 20 parts. 1788-91. Ed. by L.G.Brugnatelli. Contained articles by A. Volta and several others.

Bergbaukunde. 2 vols. 1789-90. Ed. by I. von Born* and F.W. von Trebra*. Contained articles by seven of the scientists in the survey.

Table 4

Examples of General Learned Journals Excluded from the Catalogue

France

Nouvelles de la république des lettres. 1684-1718.
Histoire des ouvrages des savans. 1687-1709.
Bibliothèque ancienne et moderne. 1714-27.
Mercure de France. 1724-91.
Nouvelle bibliothèque ou histoire littéraire. 1738-44.
Journal encyclopédique. 1756-93.
Journal littéraire. 1772-76.
Journal de Paris. 1777-1840.

Britain

Ladies' diary. 1704-1840.
Gentlemen's magazine. 1731-1907.
Gentlemen's mathematical companion. 1798-1827.

Germany

Miscellanea Lipsiensia. 1716-54.
Exercitationum subsecivarum Francofurtensium. 1717-20.
Bibliothèque germanique. 1720-41.
Halle'sche wöchentliche Anzeigen. 1729-1810.
Commercium litterarium Norimbergae. 1731-45.
Journal littéraire de l'Allemagne. 1741-44.
Altonaische gelehrte Anzeigen. 1745-89.
Hamburgisches Magazin. 1747-81.
Jenaische gelehrte Zeitungen. 1749-86.
Erlangische gelehrte Anmerckungen. 1749-89.
Hannöverische gelehrte Anzeigen. 1750-54.
Hannöverisches Magazin. 1763-90.
Lausitzisches Magazin. 1768-92.
Schlözer's Briefwechsel. 1776-82.
Hanau'sches Magazin. 1778-85.
Pfalzbayrische Beiträge zur Gelehrsamkeit. 1782-90.
Hessische Beiträge. 1785-?
Braunschweigisches Magazin. 1788-1849.

Italy

Galleria di Minerva. 1696-1717.
Giornale de' letterati d'Italia. 1710-90.
Giornale di Roma. 1716-1870.
Raccolta di opuscoli scientifici e filologici. 1728-57.
Osservazioni letterarie. (Verona) 1737-40.
Storia letteraria d'Italia. 1748-55.
Nuova raccolta d'opuscoli scientifici e filologici. 1758-87.
Giornale d'Italia. 1764-95.
Antologia di Roma. 1775-98.
Giornale enciclopedico d'Italia. 1785-89.
Biblioteca oltremontane. 1787-?

Other Countries

Nova literaria Helvetica. 1703-15.
Journal littéraire de La Haye. 1713-37.

Journal de Genève. 1772-92.
Algemeen magazijn van wetenschappen, konst en smaak. 1785-91.
American museum. 1787-92.

learned periodicals (i.e. those covering the whole range of learning) and short-lived scientific periodicals, though the distinction between the two is sometimes hard to draw.

The short-lived scientific periodicals were nearly all non-institutional. Most of them lasted for less than five years, some for only a few issues. Scientific eminence was no guarantee of success in the publishing world: there are several cases of outstanding scientists who edited periodicals which ceased after a few years, as can be seen from Table 3 which lists some of the more significant of the short-lived items excluded. There were many others, especially for natural history; the latter are hardly represented in Table 3 because of the lack of naturalists in the sample of scientists used.

A few of the general learned periodicals excluded were proceedings of relatively small provincial academies or societies such as those of Dijon and Rouen in France, Jena and Danzig in Germany, and Cortona and Brescia in Italy, but the great majority were non-institutional. The best known ones are listed in Table 4. They include many which played a prominent part in the culture of the Enlightenment, notably some of the French ones.

The results of the survey using the sample of 1752 scientists reveal a distribution pattern characteristic of the periodical literature of science at any period: a large proportion of the references concentrated in a relatively few "core" periodicals -- in the present case those in Table 2 -- with the rest of the references spread over a large number of minor periodicals -- minor in the present case either because they were marginal to science (however important the general learned journals may have been in other respects) or because they were short-lived. Thus a significant proportion of the scientific literature of the period is either scattered widely over the general periodical literature or buried in the short-lived scientific journals.

To return to the periodicals listed in Table 2. They are discussed in historical perspective in Part 3 but at this point it is useful to classify them into a number of different groups, as follows.

(a) The *Philosophical Transactions* and the *Histoire et Mémoires* of the Académie des Sciences. These were the two leading periodicals of the period and until the 1770s the only ones devoted entirely, or almost entirely, to science. (The other academies and societies covered other fields of learning as well.)

(b) The publications of the academies or societies of other important cities, together with the special case of the Academia Naturae Curiosorum (generally known as the Leopoldina Academy), the first of the German academies, which was patronised by the German Emperor and was not a city academy like the others. Beginning with the *Miscellanea* of the Berlin Academy in 1710 there are twenty-six of these academy or society publications in the list, the largest of them being those of Edinburgh, Stockholm, Turin, Berlin, St. Petersburg, and Prague.

(c) Non-institutional learned journals, beginning with the *Journal des Sçavans* which was not an important vehicle for scientific publication after 1700. Next is the *Acta Eruditorum* (04) whose importance for science also declined during the early eighteenth century. The other members of this group are the *Mémoires de Trévoux* (06), the *Göttingische Zeitung* (015), and the *Giornale dei Letterati* (08). As can be seen from the figures given for these journals in Table 2, they were quite minor as regards their scientific content, though not so minor as the large number of similar journals that were excluded. It is here that the line was drawn.

(d) The two natural history periodicals, *Der Naturforscher* (0593) and the publication of the Gesellschaft Naturforschender Freunde zu Berlin (0594). As is discussed in Part 3, they represent the beginnings of the move towards specialization in scientific periodicals.

(e) Three non-institutional journals which mark a new stage in the development of scientific periodicals. The first and by far the most outstanding of them is Rozier's Journal (030) which, as is discussed later, was the first non-institutional journal devoted entirely to the natural sciences which was able to survive for more than a few years. The other two, the *Magazin für das Neuste aus der Physik und Naturgeschichte* (035) and the *Scelta di Opuscoli Interessanti* (032) were broadly similar but less important. To these we could also add the natural history journal already mentioned, *Der Naturforscher*, which was of a similar type but partly specialized.

(f) The periodical of a general science society of a new type, the Società dei Quaranta, later known as the Società Italiana delle Scienze (037). As well as being national in scope it did not have the exclusive character that was fairly general in the eighteenth-century academies and societies (at least on the Continent). In its much wider membership it was more akin to the societies of the nineteenth century. The natural history society already mentioned, the Gesellschaft Naturforschende Freunde zu Berlin, was of a similar character but more specialized in its interests and not national in its scope.

The survey described above, using Poggendorff's bibliographical data, was concerned with periodicals covering all the natural sciences but it also yielded some information about the first specialized periodicals in non-biological fields. Those which began before 1790 are listed in Table 5 with the same kind of numerical data as are given in Table 2. As we have already seen, there were some short-lived specialized periodicals which began earlier (e.g. those in Table 3) but the ones in Table 5 were the first to survive for more than a few years. They are discussed under the headings of the various individual sciences in Section 3.2.

As an appendix to the present section it may be mentioned that Reuss' *Repertorium Commentationum a Societatibus Litterariis Editarum* can be used to sketch the outlines of the periodical literature which emanated from the academies and other learned societies during the period 1665-1800. As mentioned earlier, Reuss' compilation does not

Table 5

Specialized Periodicals Beginning Before 1790

Not including natural history periodicals. The four astronomical ephemerides which head the list were not periodicals of the usual kind but served for a time in a dual role; see Section 3.2: Astronomy.

	Period[a]	No. of Authors	Annual Average
Connaissance des temps (0351)[b]	(1766)-(1799)	17	0.50
Ephemerides astronomicae (0353)	1757-(1799)	9	0.21
Effemeridi astronomiche (0355)	1774-(1799)	4	0.15
Astronomisches Jahrbuch (0356)	1774-(1799)	61	2.54
Crell's Journal (0391)	1778-(1799)	104	4.73
Almanach für Scheidekünstler (0392)	1780-(1799)	11	0.55
Bergmannisches Journal (0454)	1788-(1799)	17	1.42
Annales de chimie (0393)	1789-(1799)	36	3.27

Notes: (a) As in Table 2 the end of the period is for present purposes taken to be 1799.
(b) The *Connaissance des temps* dates from 1679 but it did not begin to include astronomical articles until 1766.

include non-institutional periodicals, but the literature that it does include was, after all, by far the biggest and most important part of the periodical literature of the period.

The *Repertorium* is arranged by subjects, the two main groups being the natural sciences (in which group we may include mathematics) and medicine, with about six and a half volumes each. The other three groups -- "Oeconomia", "Historia", and "Philologia" -- have only one volume each. Each volume of the *Repertorium* has a table of contents which constitutes a scheme of sub-division, often quite elaborate, of the subject matter of the volume. In the natural sciences, to say nothing of the other fields, Reuss' categories and sub-categories are often very different from those of today but if, with due circumspection, we translate them into twentieth-century terms we arrive at results such as those set out in Table 6, where the number of pages for each science can serve as a rough measure of the number of papers devoted to it in the periodical literature of the period. Not surprisingly, the percentages calculated for the various sciences are mostly different from those presented in the Appendix (Table 50) for the beginning of the twentieth century, but the differences are in line with general trends in the history of the sciences. Thus astronomy was a much bigger science, and chemistry a much smaller one, in the earlier period, while on the other hand botany and zoology have much the same positions in both periods.

Table 6

Contents of the Natural Science Section of Reuss' *Repertorium*

(In Modern Terminology)

	Number of Pages	Percentage of Total
Mathematics pp. 1-168 of Vol. VII	169	6
Astronomy pp. 1-532 of Vol. V	533	19
Mechanics pp. 3-12, 51-59 of Vol. IV pp. 170-241, 243-246, 248-266 of Vol. VII	114	4
Physics pp. 60-75, 252-389 of Vol. IV	154	6
Chemistry pp. 1-190 of Vol. III	191	7
Geology (including mineralogy and palaeontology) pp. 289-566 of Vol. II pp. 14-37 of Vol. IV	301	11
Meteorology pp. 75-175 of Vol. IV	101	4
Natural History (general) pp. 1-74 of Vol. I	75	3
Botany pp. 1-286 of Vol. II	287	11
Zoology (including comparative anatomy, embryology, and physiology) pp. 75-574 of Vol. I	500	18
Human Anatomy and Physiology pp. 12-300 of Vol. X	289	11
	2714	100

2.2 The Nineteenth Century

For the nineteenth century the Royal Society's *Catalogue of Scientific Papers*, which includes both institutional and non-institutional periodicals, makes possible a more direct approach to an assessment of their relative degree of use. The Royal Society's *Catalogue* is in several series, the first (published in 1867-72) covering the period 1800-63, the second (published in 1877-79) covering the period 1864-73, and so on for the rest of the century. Each series consists of an alphabetical sequence of authors, each with the references to his periodical articles listed under his name. For the reason given below only Series 1 and 2 (together covering the period 1800-73) were used.

The method adopted for the survey used a 10% sample: all the references on every tenth page of each series were tallied under the names of the periodicals; the total number of references to each periodical was then multiplied by ten (because a 10% sample was used) and divided by the number of years from the commencement of the periodical (or, in the case of the older periodicals, from the beginning of the period covered by the *Catalogue*) up to 1863 (or 1873). The resulting figure--the average number of references per year--is used as an index of the relative size of the periodical as discussed below.

In most cases it was possible to get separate figures for the period up to 1863 (obtained from Series 1) and for the period 1864-73 (obtained from Series 2). Furthermore, in many cases the period up to 1863 could be sub-divided as a consequence of the periodical having been issued in two or more successive series; in such cases the size of the periodical during each sub-division of the period is given in Part 1, though occasionally there are gaps due to sections of the periodical not being available to the compilers of the Royal Society's *Catalogue*.

For the last years of the century the same procedure could have been used with the later series of the Royal Society's *Catalogue* but it was decided instead to use the *International Catalogue of Scientific Literature* because, unlike the Royal Society's *Catalogue*, it is subdivided by subjects. This feature made possible the compilation of the data discussed in the Appendix and also facilitated the classification of the periodicals in Part 1.

The *International Catalogue of Scientific Literature* was begun in 1901 to do for the twentieth century what the Royal Society's *Catalogue* had done for the nineteenth, but as a current rather than a retrospective periodical index; thus it was issued annually and each issue was divided into sections for the various individual sciences, each section containing both an author catalogue and a subject catalogue.* It ceased in 1917 after fourteen annual issues--a casualty of the First World War. For the short period that it covers it can be useful to the historian of science and in the present connection it serves as a source of data concerning the periodicals which existed at the beginning of the twentieth century.

---

*For information concerning the genesis and execution of this ambitious project see the Preface to the *Catalogue* and Lyons (1944, pp. 293-4, 309-11).

In the case of the *International Catalogue* a 20% sample was used (larger than before because of the extent of sub-division of the catalogue) and was spread over the first five issues--which cover the literature from 1901 to 1905--in order to smooth out year-to-year fluctuations. Thus in each of the subject sections all the references on every fifth page of the author catalogues of the first five issues were tallied under the names of the periodicals. For each periodical the figure for the average number of references per year during the period 1901-05 was then obtained from the number counted by multiplying it by five, because a 20% sample was used, and dividing it by five because the count was for five years.

The figures for the average number of references per year which were obtained from the two surveys just described were compared with direct counts of the numbers of papers in several nineteenth-century periodicals and were found to be accurate enough to indicate the relative sizes of the periodicals. There were a number of sources of possible errors: besides sampling errors there were in some cases ambiguities in the bibliographical data about the precise number of years a periodical had been in existence; furthermore in both surveys there was probably some over-counting in the case of papers with more than one author (but such papers were much less common in the nineteenth century than they are in the twentieth).

Indications of the sizes of the periodicals during various periods of time are given in Part 1 using the scale set out in Table 7 in which each category of size represents a range of average numbers of references per year.

Table 7

Categories of Size

Category	References per Year	Category	References per Year
1	-9	11	100-199
2	10-19	12	200-299
3	20-29	13	300-399
4	30-39	14	400-499
5	40-49	15	500-599
6	50-59	16	600-699
7	60-69	17	700-799
8	70-79	18	800-899
9	80-89	19	900-999
10	90-99	20	1000-

Periodicals were included in Part 1 if their average number of references per year for any period within the nineteenth century, or the period 1901-05, was above a certain minimum--generally three or four, depending on various other considerations. Bibliographical periodicals, for which there were normally no references, were always included. An average of three or four references per year might seem to indicate a very small periodical but it should be remembered that nineteenth-century periodicals commonly contained other material (news items, personalia, book reviews, etc.) besides original articles. Moreover, the periodicals issued by the many small general science societies commonly contained a substantial proportion of articles which were not counted because they were on topics (technological, anthropological, sociological, etc.) outside the subject range of the catalogues used or of the present work, or both.

A very large number of periodicals was excluded, mainly short-lived items and the publications of small local societies. There were also some general learned periodicals but these were much less prominent than they had been in the eighteenth century. Other categories of excluded material are mentioned in Part 3.

It is useful to rank the periodicals in order of size and this has been done at two points--or short periods--in time, corresponding to the two surveys described above. For the beginning of the twentieth century the data from the second survey, using the *International Catalogue* in the period 1901-05, were used. With the data from the first survey, using Series 1 and 2 of the Royal Society's *Catalogue*, the best point in time was 1863-64, at the junction of the two series; data from both series could then be used (in most cases the mean was taken of the figures for the periods just before 1863 and just after 1864).

The results are set out in two series of tables -- Tables 8 to 19 for the period 1863-64 (up to p. 107) and Tables 20 to 33 for the period 1901-05 (pp. 107-114). The titles of the larger periodicals are given in the tables but generally only the entry numbers of the smaller ones. Periodicals which ceased before 1863 do not of course appear in any of the tables, and those which ceased in the interval 1865-1900 do not appear in the second series of tables.

Table 8

Ranking by Size, 1863-64. Science in General

Size	Periodicals
15	Académie des Sciences, *Comptes rendus* (098)
11	*The philosophical magazine* (050) *The American journal of science* (069) British Assoc. Adv. of Science, *Report* (092) [Vienna Academy], *Sitzungsberichte* (0135)
9	[Berlin Academy], *Monatsberichte* (0103) [Bonn Society], *Sitzungsberichte* (0158)
8	*Zeitschrift für Naturwissenschaften* (0154) [Naples Academy], *Rendiconto* (0175) *Les mondes: Revue hebdomadaire* (0179)
7	Royal Society, *Proceedings* (094)
6	[Brussels Academy], *Bulletins* (093) [Wurzburg Society], *Verhandlungen* (0143)
5	Physikal. Verein, Frankfurt, *Jahresbericht* (0104)
4	061, 071, 0101, 0112, 0118, 0129, 0136, 0162.
3	02, 086, 095, 0102, 0106, 0120, 0124, 0131, 0145, 0166.
2	044, 055, 058, 065, 066, 096, 097, 099, 0106. 0109, 0114, 0121, 0125, 0126, 0130, 0132, 0133, 0134, 0140, 0141, 0148, 0149, 0152, 0153b, 0155, 0159, 0163, 0164, 0167, 0171, 0173, 0177, 0180, 0181.

Size 1 03, 011, 014, 015, 023, 026, 029, 031, 037, 039, 040, 041, 042, 043, 046, 052, 053, 054, 056, 059, 063, 072, 073, 075, 077, 078, 080, 082, 085, 087, 088, 089, 090, 0100, 0105, 0117, 0119, 0123, 0127, 0128, 0137, 0138, 0139, 0142, 0144, 0146, 0150, 0151, 0153a, 0156, 0157, 0160, 0161, 0165, 0168, 0169, 0170, 0172, 0174, 0176.

Note A The Royal Society's *Catalogue* has no references to the following items in or near the period 1863-64 (presumably because they were unavailable). Some or all of them may have been fairly large.
Soc. Philomat. de Paris, *Bulletin* (045)
Gesellsch. Deutsch. Naturf. Ärzte, *Bericht* (079)
Skand. Naturf. Läkare, *Fördlandlingar* (0107)

Note B The Royal Society's *Catalogue* does not include the following semi-popular items (which are included in the *International Catalogue*).
Scientific American 0122)
Die Natur (0147)

Table 9

Ranking by Size, 1863-64. Mathematics

Size 6 *Archiv Math. und Physik* (Ed. Grunert) (0282)

5 *Nouv. annales de math.* (0283)
Giornale di matematiche (0288)

4 *Journal de math.* (Ed. Liouville) (0280)
Quarterly journal pure and appl. math. (0281)
Zeitschrift für Math. und Physik (0285)

3 *Journal für reine und angew. Math.* (Ed. Crelle) (0279)
The messenger of math. (0287)

2 *Annali di matematica* (0284)

1 *Tidsskrift for math.* (0286)

Table 10

Ranking by Size, 1863-64. Astronomy

Size 9 *Astronomische Nachrichten* (0329)

5 Royal Astronom. Soc., *Monthly notices* (0330)

1 Royal Astronom. Soc., *Memoirs* (0328)
Collegio Romano, Osserv., *Memorie* (0332)
Harvard College, Astronom. Observ., *Annals* (0333)

Table 11

Ranking by Size, 1863-64. Physics

Size 11 *Annalen der Physik* (0362)

6 *Il nuovo cimento* (0365)

2 *Annales télégraphiques* (0367)

Table 12

Ranking by Size, 1863-64. Chemistry

Size 12	*Journal für praktische Chemie* (0407)
11	*Annales de chymie* (0393)
	Journal de pharmacie et de chimie (0399)
	Annalen der Chemie (Ed. Liebig) (0409)
	Chemical news (0419)
10	*Zeitschrift für Chemie* (0418)
9	Chemical Society, London, *Journal* (0412)
7	Société Chimique de Paris, *Bulletin* (0417)
5	*Pharmaceutical journal* (0413)
4	*Annali di chimica* (0404c)
	Zeitschrift für analytische Chemie (0420)
3	*Archiv der Pharmazie* (0402)
	Chemisches Central-Blatt (0408)
	Moniteur scientific (0410)
2	*Journal de chimie médicale* (0405)
1	*Repertorium für die Pharmacie* (0400)
	American journal of pharmacy (0406)

Table 13

Ranking by Size, 1863-64. Geology

Size 7	Geological Society of London, *Quarterly Journal* (0462)
6	Société Géologique de France, *Bulletin* (0464)
5	*Neues Jahrbuch für Mineralogie....* (0456b)
	Deutsche Geolog. Gesellsch., *Zeitschrift* (0474)
4	Geolog. Reichanstalt, Wien, *Jahrbuch* (0475)
3	*Annales des mines* (0455)
2	0458, 0471, 0473, 0478, 0479, 0481, 0482a, 0484.
1	0459, 0461, 0463, 0465, 0466, 0467, 0468, 0469, 0472, 0476, 0477, 0480, 0483, 0486.

Table 14

Ranking by Size, 1863-64. Geography

Size 3	Royal Geographical Society, London, *Journal* (0541)
	Geographische Gesellschaft, Wien, *Mittheilungen* (0552)
2	Société de Géographie, Paris, *Bulletin* (0540)
	American Geographical Society, *Bulletin* (0547)
	Royal Geographical Society, London, *Proceedings* (0550)
	Justus Perthes' Geogr. Anstalt, *Mittheilungen*, (0551)
1	0543, 0544, 0545, 0546, 0548, 0549.

Table 15

Ranking by Size, 1863-64. Meteorology

Size 4 Scottish Meteorological Society, *Report* (0576)

3 British Meteorological Society, *Proceedings* (0577)

2 Collegio Romano, Osserv., *Bulletino meteorol.* (0578)

Table 16

Ranking by Size, 1863-64. Natural History

Size 11 *Annals and mag. of nat. hist.* (0614)

8 *Natuurkund. tijdschr. voor Ned.-Indië*, (0630)

6 Société Linnéenne de Lyon, *Annales* (0604)

5 Acad. of Nat. Sci., Philadelphia, *Proceedings* (0616)
Zool.-Bot. Gesellsch., Wien, *Verhandlungen* (0633)

4 Soc. Imp. Naturalistes, Moscou, *Bulletin* (0606)
Boston Soc. of Nat. Hist., *Proceedings* (0617)

2 0598, 0618, 0621a, 0621c, 0625, 0628, 0631, 0634, 0636, 0638, 0641.

1 0595, 0596, 0599, 0601, 0603, 0607, 0609, 0610, 0611, 0612, 0613, 0619, 0620, 0621b, 0623, 0624, 0626, 0627, 0629, 0632, 0635, 0637, 0639, 0640.

Table 17

Ranking by Size, 1863-64. Botany

Size 8 *The journal of botany* (0713)

5 *Botanische Zeitung* (0698)
Société Botanique de France, *Bulletin* (0706)

4 *Österreich. botanische Zeitschrift* (0702)
Adansonia: Recueil périod. obser. bot. (0710)

3 *Flora* (0685)
Botanical Soc. of Edinburgh, *Transactions* (0692)
Botaniska notiser (0693)

2 0689, 0695, 0707, 0708, 0712.

1 0687, 0701, 0703, 0704, 0709, 0711.

Table 18

Ranking by Size, 1863-64. Zoology

Size 9 Société Entomol. de France, *Annales* (0806)

8 Zoological Society , London, *Proceedings* (0756)

6 *Berliner entomologische Zeitschrift* (0812)

5 *The zoologist* (0761)

4 0764, 0769, 0809, 0811, 0837, 0838, 0854.

3 0758, 0763, 0815.

2 0757, 0759, 0805, 0807, 0813, 0814, 0853.

1 0753, 0762, 0765, 0766, 0767, 0770, 0810.

Table 19

Ranking by Size, 1863-64. Experimental Biology

Size 6 *Archiv für Anat., Physiol....* (Ed. Müller) (0872a)

5 Société de Biologie, *Comptes rendus* (0876)

4 *Archiv für path. Anat. u. Physiol.* (Ed. Virchow) (0874)

2 *Untersuch. zur Naturlehre Menschen u. Thiere* (Ed. Moleschott) (0878)
Journal physiolog. homme et animaux (0881)

1 0871, 0875, 0879, 0880.

Table 20

Ranking by Size, 1901-05. Science in General

Size 20 Académie des Sciences, *Comptes rendus* (098)

12 *Science* (0244)

11 *The philosophical magazine* (050)
The American journal of science (069)
Gesellsch. Deutsch. Naturf. u. Ärzte, *Bericht* (079)
British Assoc. Adv. of Science, *Report* (092)
Royal Society, *Proceedings* (094)
[Berlin Academy], *Sitzungsberichte* (0103)
Archives des sciences phys. et nat. (0129)
[Vienna Academy], *Sitzungsberichte* (0135)
Die Natur (0147)
Science gossip (0192)
Nature [London] (0206)
Accademia dei Lincei, *Rendiconti* (0211)
[Amsterdam Academy], *Verslagen* (0266)

10 Assoc. Française Adv. des Sci., *Comptes rendus* (0218)
La nature (0220)

9 *Naturwissenschaftliche Wochenschrift* (0251)

7 *Scientific American* (0122)
[Cracow Academy], *Rosprawy* (0224, 0259)
Revue générale des sciences (0260)

6 *Philosophical transactions* (02)
St.-Peterburg. Obshchest. Estest. *Trudui* (0214)
Popular science monthly (0217)

5 0120, 0185, 0189, 0193, 0197, 0201, 0205, 0250.

4 086, 097, 0101, 0112, 0124, 0125, 0127, 0128, 0179, 0180, 0194, 0198, 0207, 0240, 0264.

3 061, 0118, 0126, 0154, 0158, 0161, 0175, 0176, 0187, 0191, 0216, 0227, 0235, 0236, 0238, 0239, 0254, 0256, 0257, 0262, 0270, 0274.

2 014, 039, 044, 055, 059, 075, 080, 085, 0104, 0106, 0119, 0140, 0141, 0143, 0152, 0162, 0163, 0164, 0170, 0172, 0173, 0181, 0190, 0199, 0202, 0204, 0209, 0212, 0226, 0229, 0231, 0232, 0237, 0249, 0253, 0261, 0268, 0272, 0273, 0276.

1 03, 011, 015, 023, 026, 029, 031, 037, 041, 042, 043, 045, 046, 052, 053, 054, 056, 058, 063, 065, 066, 072, 073, 077, 078, 082, 087, 088, 089, 090, 093, 099, 0100, 0102, 0105, 0109, 0114, 0116, 0117, 0121, 0123, 0130, 0131, 0132, 0133, 0137, 0138, 0139, 0142, 0145, 0146, 0148, 0150, 0151, 0153a, 0156, 0157, 0159, 0160, 0166, 0167, 0168, 0169, 0171, 0174, 0177, 0178, 0182, 0184, 0186, 0188, 0195, 0196, 0200, 0203, 0208, 0210, 0213, 0215, 0219, 0221, 0222, 0223, 0225, 0228, 0230, 0233, 0234, 0241, 0242, 0243, 0245, 0246, 0247, 0252, 0255, 0258, 0263, 0265, 0267, 0269, 0271, 0275.

Table 21

Ranking by Size, 1901-05. Mathematics

Size 6 *Zeitschrift für Math. und Physik* (0285)

5 *Archiv der Math. und Physik* (0282)
Nouv. annales de math. (0283)
Mathematische Annalen (0294)

4 *The messenger of mathemetics* (0287)
London Mathematical Society, *Proceedings* (0289)
Zeitschrift für math. u. naturwiss. Unterricht (0295)
Société Mathématique de France, *Bulletin* (0297)
Periodico di scienze matemat. per l'insegn. sec. (0300)
Deutsche Math. Vereinigung, *Jahresbericht* (0313)
American math. monthly (0320)
Enseignement mathématique (0325)

3 *Journal für die reine und angew. Math.* (0279)
Quarterly journal of pure and appl. math. (0281)
Bibliotheca mathematica (0309)
Monatshefte für Math. und Physik (0314)
Math. és physikai lapok (0318)

2 0288, 0290, 0301, 0302, 0303, 0305, 0307, 0310, 0311, 0312, 0317, 0321, 0322, 0323.

1 0280, 0284, 0286, 0296, 0298, 0299, 0306, 0308, 0315, 0316.

Table 22

Ranking by Size, 1901-05. Astronomy

Size 14 *Astronomische Nachrichten* (0329)

10 Royal Astronom. Soc., *Monthly notices* (0330)

8 *The astrophysical journal* (0339)
Popular astronomy (0350)

6 *The astronomical journal* (0331)

5 *The observatory* (0337)
British Astronom. Assoc., *Journal* (0345)

4 *Bulletin astronomique* (0340)

3 Soc. Spettroscopisti Italiani, *Memorie* (0336)
Soc. Astronom. de France, *Bulletin* (0341)

2 0334, 0342, 0343.

1 0328, 0332, 0333, 0335, 0338, 0346, 0347, 0348, 0349.

Table 23

Ranking by Size, 1901-05. Physics

Size 12 *Annalen der Physik* (0362)
Physikalische Zeitschrift (0390)

11 *Elektrotechnische Zeitschrift* (0377)
L'électricien (0379)

9 *Zeitschrift für physik. u. chem. Unterricht* (0385)

7 *La lumière électrique* (0375)
The physical review (0388)

6 *Journal de physique théor. et appl.* (0370)
Physikal. Gesellsch., Berlin, *Verhandlungen* (0381)

5 *Zeitschrift für Instrumentenkunde* (0380)

4 *Central-Zeitung für Optik und Mechanik* (0376)

3 0365, 0371, 0384, 0387.

2 0369, 0374, 0386.

1 0372, 0378, 0382, 0383.

Table 24

Ranking by Size, 1901-05. Chemistry

Size 17 Deutsche Chemische Gesellschaft, *Berichte* (0421)

12 Société Chimique de Paris, *Bulletin* (0417)
Russ. Fiziko-Khim. Obshchest., *Zhurnal* (0422)
Zeitschrift für physiologische Chemie (0427)
Chemiker-Zeitung (0428)

11 Chemical Society, London, *Journal* (0412)
American Chemical Society, *Journal* (0430)
Zeitschrift für physikalische Chemie (0437)
Zeitschrift für angewandte Chemie (0439)
Zeitschrift für anorganische Chemie (0440)
Zeitschrift für Elektrochemie (0444)

10 *Zeitschrift für Untersuch. Nahrungs- u. Genussmittel* (0436)

9 *Justus Liebig's Annalen der Chemie* (0409)
Society of Chemical Industry, *Journal* (0433)

8 *Monatshefte für Chemie* (0431)

7 *Annales de chimie* (0393)
Archiv der Pharmazie (0402)
Zeitschrift für analytische Chemie (0420)
Chemical Society, London, *Proceedings* (0435)
Annales de chimie analytique (0451)

6 0407, 0419, 0424, 0429.

5 0447, 0450.

4 0399, 0434, 0445.

3 0410, 0413, 0426, 0438, 0448, 0449.

2 0425, 0432, 0452.

1 0406.

Table 25

Ranking by Size, 1901-05. Geology

Size 11 *The geological magazine* (0482)

9 Société Géologique de France, *Bulletin* (0464)
Geological Survey of the United States, [Publications] (0504)

8 Geological Society, London, *Quarterly journal* (0462)

7 Deutsche Geologische Gesellschaft, *Zeitschrift* (0474)
Zeitschrift für Krystallog. u. Mineralog. (0501)
The American geologist (0516)

6 Geologische Reichanstalt, Wien, *Verhandlungen* (0481)

5 *Neues Jahrbuch für Mineralogie....* (0456)
Geological Society of America, *Bulletin* (0522)
Zeitschrift für praktische Geologie (0526)

4 Geologische Reichanstalt, Wien, *Jahrbuch* (0475)
Tschermaks mineralog. u. petrograph. Mittheilungen (0494)
Internat. Geological Congress, *Proceedings* (0502)
Società Geologica Italiana, *Bollettino* (0509)
Journal of geology (0525)

3 0471, 0486, 0495, 0505, 0510, 0519, 0527.

2 0465, 0468, 0469, 0483, 0492, 0493, 0499, 0500, 0503, 0507, 0514, 0523, 0524, 0528, 0530, 0532, 0535, 0536, 0537, 0539.

1 0455, 0458, 0459, 0461, 0463, 0466, 0472, 0473, 0476, 0477, 0478, 0479, 0480, 0484, 0487, 0489, 0490, 0491, 0496, 0497, 0506, 0508, 0511, 0512, 0513, 0515, 0517, 0518, 0520, 0521, 0529, 0531, 0534, 0538.

Table 26

Ranking by Size, 1901-05. Geography

Size 9 *The geographical journal* (0550)

7 *Petermanns geogr. Mittheilungen aus Perthes' geogr. Anstalt* (0551)

5 *Globus* (0553)

4 Société de Géographie, Paris, *Bulletin* (0540)
Internat. Geographical Congress, [Publications] (0557)

3 Gesellschaft für Erdkunde, Berlin, *Zeitschrift* (0548)
Scottish geographical magazine (0564)
Geographische Zeitschrift (0569)

2 0552, 0554, 0556, 0559, 0562, 0563, 0565, 0566, 0570.

1 0543, 0546, 0547, 0549, 0555, 0558, 0560, 0561, 0567, 0568, 0571.

Table 27

Ranking by Size, 1901-05. Meteorology

Size 11 *Meteorologische Zeitschrift* (0582)

8 *Monthly weather review* (0584)

6 *The meteorological magazine* (0579)
Das Wetter (0583)

3 0575, 0577, 0585.

1 0576, 0581.

Table 28

Ranking by Size, 1901-05. Natural History

Size 11 *Annals and mag. of natural history* (0614)

9 *The naturalist* (0632)
Natur und Haus (0671)

7 Academy of Nat. Sci., Philadelphia, *Proceedings* (0616)
The American naturalist (0642)
The Irish naturalist (0672)

6 *Annals of Scottish natural history* (0647b)
Bombay Natural History Society, *Journal* (0667)

5 *Feuille des jeunes naturalistes* (0646)
Természetrajzi füzetek (0654)

4 Société Linnéenne de Bordeaux, *Bulletin* (0603)
Société Vaudoise des Sci. Nat., *Bulletin* (0618)
Naturhist. Forening, Kjøbenhavn, *Videnskab. meddelelser* (0628)
Zool.-Bot. Gesellsch., Wien, *Verhandlungen* (0633)
Gesellsch. Naturf. Freunde, Berlin, *Sitzungsber.* (0639)
Linnean Society of N.S.W., *Proceedings* (0652)
The Victorian naturalist (0666)

3 Société Imp. des Naturalistes, Moscou, *Bulletin* (0606)

2 0619, 0621a, 0623, 0625, 0635, 0636, 0641, 0643, 0644, 0650, 0651, 0653, 0665.

1 0595, 0596, 0598, 0601, 0604, 0608, 0609, 0610, 0612, 0613, 0617, 0620, 0624, 0626, 0629, 0630, 0631, 0637, 0638, 0640, 0645, 0648, 0649, 0655, 0656, 0658, 0659, 0662, 0663, 0664, 0668, 0669, 0670, 0673.

Table 29

Ranking by Size, 1901-05. Microscopy

Size 7 *Zeitschrift für wissensch. Mikroscopie* (0679)

2 *Quarterly journal of microscopical science* (0675)
Quekett Microscopical Club, London, *Journal* (0676)
Zeitschrift für angewandte Mikroskopie (0680)

1 0674, 0677, 0678.

Table 30

Ranking by Size, 1901-05. Botany

Size 11 *The journal of botany* (0713)

10 *The gardener's chronicle* (0695)
Deutsche Botanische Gesellsch., Berlin, *Berichte* (0728)

9 *Rhodora* (0751)

7 *Die Gartenflora* (0704)
Botanisches Centralblatt (0725)

6 Torrey Botanical Club, *Bulletin* (0716)
The botanical gazette (0723)

5 Société Botanique de France, *Bulletin* (0706)
Botanische Jahrbücher.... (0726)
The fern bulletin (0743)
Allgemeine botanische Zeitschrift.... (0747)

4 0685, 0708, 0715, 0732, 0738, 0741, 0749.

3 0693, 0698, 0700, 0702, 0703, 0709, 0718,
0727, 0735, 0736, 0740, 0744, 0750.

2 0692, 0699, 0707, 0719, 0722, 0729, 0730,
0733, 0734, 0742, 0748.

1 0689, 0701, 0711, 0712, 0717, 0724, 0737,
0739, 0745, 0746.

Table 31

Ranking by Size, 1901-05. Zoology

Size 11 Zoological Society, London, *Proceedings* (0756)
The zoologist (0761)
Zoologischer Anzeiger (0776)

10 Muséum National d'Histoire Naturelle, *Bulletin* (0800)

8 *Zoologische Jahrbücher* (0785)
The auk: Quarterly journal of ornithology (0841)
Ornithologische Monatsberichte (0850)

7 Société Entomologique de France, *Bulletin* (0834)

6 *Zeitschrift für wissenschaftliche Zoologie* (0763)
Der zoologische Garten (0769)
Le naturaliste (0777)
The entomologist's monthly magazine (0816)
Entomological news (0831)

Allgemeine Zeitschrift für Entomologie (0835)
Ornithologische Monatsschrift (0842)
British Ornithologists' Club, *Bulletin* (0848)
The nautilus (0858)

5 0772, 0792, 0807, 0829, 0838.

4 0759, 0779, 0797, 0814, 0823, 0827, 0860, 0868.

3 0770, 0778, 0786, 0791, 0801, 0805, 0813, 0824, 0828, 0833, 0837, 0839, 0851, 0862, 0866.

2 0774, 0780, 0781, 0782, 0783, 0784, 0793, 0794, 0795, 0796, 0806, 0809, 0812, 0819, 0821, 0822, 0825, 0826, 0830, 0832, 0840, 0849, 0854, 0856, 0857, 0859, 0864.

1 0753, 0757, 0758, 0762, 0765, 0766, 0767, 0773, 0775, 0787, 0788, 0790, 0798, 0799, 0803, 0804, 0811, 0815, 0817, 0818, 0820, 0843, 0845, 0846, 0847, 0852, 0855, 0861, 0863, 0865, 0867.

Table 32

Ranking by Size, 1901-05. Experimental Biology

Size 16 Société de Biologie, Paris, *Comptes rendus* (0876)

13 *Anatomischer Anzeiger* (0903)

12 *Archiv für Anatomie und Physiologie* (0872 b,c)
Archiv für die gesammte Physiologie (0889)

11 *Archiv für path. Anatomie und Physiologie* (0874)
Archiv für mikroskopische Anatomie (0883)
Archives italiennes de biologie (0898)

10 *Biologisches Centralblatt* (0897)
Anatomische Hefte (0911)

9 *Centralblatt für allgemeine Pathologie* (0909)
Archiv für Entwicklungsmechanik der Organismen (0916)

8 *American journal of physiology* (0922)

7 *Journal of physiology* (0893)
Beiträge zur pathologischen Anatomie (0901)

5 Société Anatomique, Paris, *Bulletins* (0871)
Journal of anatomy (0887)
Archiv für exper. Pathologie und Pharmakologie (0891)
Biological Society, Washington, *Proceedings* (0895)

4 *Zeitschrift für Biologie* (0884)

3 0879, 0892, 0894, 0899, 0900, 0907, 0912, 0913, 0914, 0923.

2 0875, 0881, 0882, 0904, 0910, 0920, 0921.

1 0878, 0890, 0896, 0902.

Table 33

Ranking by Size, 1901-05. Microbiology

Size	
15	*Centralblatt für Bakteriologie u. Parasitenkunde* (0930)
8	*Zeitschrift für Hygiene* (0928)
4	*Archiv für Hygiene* (0925) Institut Pasteur, *Annales* (0929) *Russky arkhiv patol., klin. med. i bakteriol.* (0935)
3	*Archives de parasitologie* (0936)
2	*Journal of pathology and bacteriology* (0933)
1	0924, 0931, 0934.

PART 3

THE DEVELOPMENT OF THE PERIODICAL LITERATURE*

3.1 Science in General

The vast importance of the initiation of periodical publication by the *Journal des Sçavans* on 5 January 1665 is universally recognized. Though it included natural science the *Journal des Sçavans* cannot be called a scientific periodical, rather it was the first member of the genre of general learned journals which was to become a characteristic feature of the Age of Enlightenment. The *Philosophical Transactions*, begun only two months later, was sufficiently different in nature to warrant the title of the first scientific periodical. It was launched by Henry Oldenburg, the indefatigable secretary of the Royal Society which had come into existence a few years earlier, and it closely reflected the concerns of the Society.

The *Philosophical Transactions* was an immediate success not only in Britain but also on the Continent even though English was not then a well-known language. "In spite of its English dress, and the difficulty of conveying packets over long distances (for the *Transactions* could not go by ordinary post) the journal was welcomed abroad, and scientists everywhere begged to have copies sent by sailors, merchants, the diplomatic bag or chance travellers." (Hall, 1975) Further dissemination of its contents resulted from the journalistic practice, common then and for a long time afterwards, of borrowing from other periodicals: the *Journal des Sçavans* and the similar journals which followed it often contained translations of extracts from it (and in the earlier volumes Oldenburg often printed English versions of extracts from them).

Yet it was not until 1753 that the Royal Society officially acknowledged the *Transactions* as its own publication. Until then it both was and was not an institutional journal. It was *not* because it had been begun as a private venture by Oldenburg and was continued as such by his successors, and because the Society, perhaps wary of being compromised, explicitly disclaimed responsibility for it on more than one occasion. And it *was* an institutional periodical because, as well as being printed by the Society's printers and under its licence, it was intimately connected with the Society in many ways and derived

*This survey is based on the data in Part 1 together with the published literature dealing with the subject; numerous references to the literature are given in Part 1 and are mostly not repeated here. The account also draws on the biographies of editors, the history of scientific societies, and the social history of science generally.

most of its contents from the members' activities -- and because everyone thought it was, at least everyone on the Continent.

The *Journal des Sçavans*, too, was a success in its wider field and it was not long before a number of similar publications appeared in several countries, some of them short-lived. The most notable periodical of this type in the late seventeenth century was the *Acta Eruditorum* (04) published monthly at Leipzig, the centre of the German book trade, from 1682. Law, theology, and medicine figured prominently in its pages as well as natural science and, more than its counterparts in western Europe, it had a transitional character, incorporating the scholasticism of the past as well as the new learning. "A rapid perusal of the first volumes however shows that the *Acta* were in sympathy with the new science. The first volume opens with Nehemiah Grew's work; Boyle, Sydenham, Papin, Borelli, Leeuwenhoek, Leibniz, Sturm, Bernoulli, Hevelius, were represented in its pages, and the following volumes continued to count the most progressive scientists among their contributors." (Ornstein, 1938) At a time when Germany, still suffering from the consequences of the Thirty Years War, lagged behind western Europe, the *Acta Eruditorum*, "written by scholars for scholars", played a most valuable role as a focus for German scholarship and a means of contact with developments elsewhere.

There were no close imitations of the *Philosophical Transactions* but it was acknowledged to have provided much of the inspiration for what can be regarded as the second scientific periodical (though it was also medical), the *Miscellanea Curiosa Medico-Physica* (03), begun in 1670 as the organ of a small German society which was later enlarged and adopted by the Emperor, becoming in 1682 the Academia Caesareo-Leopoldina Naturae Curiosorum, the scientific (including medical) academy of the Holy Roman Empire. Its periodical was quite substantial until the early eighteenth century but thereafter it shrank to a small size, no doubt because the Academy was affected by the decline of the Empire and its institutions, and lost ground to the other German academies which came into existence.

By the late seventeenth century France had become the leading nation of Europe, culturally as well as politically. The Académie des Sciences, established in 1666, was one of a number of royal academies which together constituted a set of cultural institutions which no other nation ever equalled. It was not until the end of the century however that the Paris Academy (not to be confused with the Académie Française, the literary academy) began a periodical. The reason for the delay was that the Academy at its foundation had adopted a policy of collective research, with the results of the research to be published--in book form--in the name of the Academy, as had recently been done by the Accademia del Cimento of Florence. A generation later this policy was seen to have been a mistake, though it was an understandable one in the earliest stage of development of scientific institutions.

Until the end of the century, then, the only scientific periodicals (apart from one or two short-lived items) were the *Philosophical Transactions* and the Leopoldina Academy's periodical, though, as has been mentioned, much science was included during this period in the *Journal des Sçavans*, the *Acta Eruditorum*, and some of the other general learned journals. French scientists especially made use of them because the Academy's policy of collective research soon broke down. Science is indeed a collective enterprise, though on a much larger scale than the early academies envisaged, but it is an individual enterprise as well. And, compared to book publication, the relatively short accounts of particular pieces of research, which periodicals can publish, are well suited to the conduct of scientific work. Consequently, until the end of the century the French academicians published their researches outside the ambit of the Academy, in the periodicals that were available to them.

The general learned journals continued to serve as significant vehicles for scientific publication until well into the eighteenth century. Those most used by the scientific community have been included in the Catalogue (01, 04, 06, 08, 015) because they could hardly be omitted, even though they were not scientific periodicals. They were especially suitable for making general announcements, such as Lalande's invitation in the *Mémoires de Trévoux* to all interested persons to participate in observing the return of Halley's comet in 1759 (Kronick, 1976, p. 255), and for carrying on controversies such as the one between Clairaut and d'Alembert in 1759-62 which involved several journals (DSB--article on Clairaut). Their various roles in relation to the science of the time have been discussed by Kronick (1976, Chapter XII).

Scientists' use of the general learned journals decreased in the course of the eighteenth century not only because the number of more suitable periodicals gradually increased but also through a process of specialization. We shall see later how in the nineteenth century the general science periodicals gradually gave way to specialized periodicals catering for the individual sciences (astronomy, chemistry, etc.). Here, a century earlier, the general science periodicals were gradually becoming distinct from the general learned periodicals. Hitherto the natural sciences had been part of general culture and it had been taken for granted by everyone that they should be included in the scholarly reviews along with other branches of learning. But very gradually science became differentiated -- the more so the more technical it became -- and grew aware of a need of its own media of publication.

A new chapter in the history of scientific periodicals began in 1699 when, as part of a thoroughgoing re-organization, the Académie des Sciences explicitly abandoned its earlier policy of collective research and initiated its *Histoire et Mémoires* (05) which immediately became the most prestigious vehicle of scientific publication, a position it continued to retain throughout the eighteenth century. Preceding the main section of the periodical, which contained the formally presented memoirs, there was a smaller section with separate pagination entitled *Histoire* which contained news of the activities of the Academy and synoptic accounts of the topics dealt with in the memoirs. The *Histoire* was addressed, in the spirit of the Enlightenment, to the educated public and was the special province of the Academy's brillant secretary, Fontenelle, the first great popularizer of science. It also included *éloges* of deceased academicians, a genre which Fontenelle created and which became remarkably popular.

The French model, in science as in other spheres, was widely imitated during the eighteenth century as enlightened monarchs set up their own national or state academies. But it was only France that had a set of specialized academies: the other Continental academies (and their periodicals) covered the whole range of learning, including philology, history, medicine, technology, agriculture, etc., as well as the natural sciences, though the latter were generally very prominent. Prussia was the first state to follow the new trend and the Berlin Academy was set up in 1700, largely inspired by Leibniz. Its *Miscellanea* (07) began in 1710 but the Academy had many difficulties in its early years and its main period dates from 1745 when it became a favorite of the new king, Frederick II, who deserves the title "the Great" at least as much for his enlightened patronage of the arts and sciences as for his military conquests. The Academy was completely re-organized, its official language changed from Latin to French, and a new series of its periodical instituted with the title *Histoire et Mémoires* (017).

During the first two-thirds of the century periodicals were begun by academies or societies in Uppsala (011), St. Petersburg (012), Bologna (013), Stockholm (014), Copenhagen (016), Basel (019), Göttingen (020), Harlem (023), Erfurt (024), Turin (025, 038), Siena (026), Munich (027), and Mannheim (028), to go no further. With the inclusion of the earlier foundations in London, Paris, and Berlin, the list comprises many of the chief cultural centres of eighteenth-century Europe, and there were also periodicals emanating from academies or societies in such provincial centres as Bordeaux (09), with Philadelphia (029) representing the New World. Most of the provincial items were not included in the Catalogue because of their smallness.

Up to 1770 it was the periodicals of the academies and similar societies which provided the science of the time with its essential means of communication. As discussed in Section 2.1 (pp. 95-97), some science was scattered through the general learned journals and some was consigned to obscurity in the short-lived scientific or semi-scientific journals, but these areas were quite minor in both quantity and quality when compared with the publications of the learned societies. For all their virtues, however, the latter had their deficiencies. A learned society, and especially an academy of the eighteenth-century type, tended to be an inward-looking institution, regarding its periodical more as an archive or repository for scientific knowledge than a means of disseminating new discoveries to the outside world. Characteristically, the periodicals of the learned societies were issued annually, or even more infrequently in many cases (the smaller societies generally waited until they had accumulated enough material to fill a new volume). In contrast, the short-lived productions of the journalists came out often, commonly monthly, and were concerned above all with announcement of the latest news and discoveries.

However suitable the societies' tomes were as archives they were far from satisfactory in their other function as means of dissemination. Moreover they were expensive, and this was a time when public libraries were very few (and university libraries, such as they were, did not count). It was their inaccessibility as well as their increasing numbers that led to the emergence of secondary or bibliographical publications, notably the *Commentarii de Rebus in Scientia Naturali* (021) and the *Collection Académique* (022), which tried to present a comprehensive view of the periodical literature of science.

It was bad enough that the societies' publications only came out annually, or even less often, but what was worse was that most of them were perpetually in arrears, sometimes by several years. A paper presented to a learned society would not appear in print for at least a year, commonly two or three years, and not infrequently longer. The august *Mémoires* of the Paris Academy was one of the worst offenders, being chronically in arrears, sometimes by as much as six years; the *Philosophical Transactions* however had a much better record. Scientists in the eighteenth century were no less anxious for speedy publication of their results than they are in the twentieth century but this was a situation which for a long time they had to accept because there was no real alternative.

In the 1770s there was a distinct quickening of tempo in the scientific world, at least in France, Germany, and Italy (British science being still in its eighteenth-century decline). The number of scientists had been increasing steadily for over a century and by the 1770s the number had become large enough to generate something like a "critical mass" effect. On the other hand the scientific institutions and periodicals had become inadequate. Academies such as those of Paris and Berlin had a strictly limited number of members (who were paid servants of the state) and for decades there had been an increasing number of able scientists who had no chance of becoming members, and consequently no chance of publishing in the academies' periodicals which were for members only. It was in consequence of this situation that the Paris Academy in 1750 had begun its series (018) for *savans étrangers*, i.e. non-members, but this amounted to not much more than a gesture, especially as it ceased in 1786, presumably for financial reasons.

In Paris in 1771 the Abbé Rozier began a new scientific periodical (030) which soon became remarkably successful and, as we shall see, was shortly followed by others of a related kind. Historians have recognized that Rozier's Journal, as it is called, marked a new stage in the development of scientific periodicals. Rozier's competent editing and his awareness of the tendencies and needs of the day were a necessary condition for its success, but no less necessary were an adequate number of contributors and readers. The time was ripe. Rozier's Journal was the first non-institutional periodical devoted entirely to original work in the natural sciences which was able to survive and even to flourish. It was issued monthly and was open to anyone, in France or elsewhere, who had research results of a proper standard to publish. As a result it became within a few years bigger than either the Paris Academy's *Mémoires* or the *Philosophical Transactions*.

The journal's contents were confined very largely to what were called at the time *les sciences physiques* which in the eighteenth century meant the experimental and inductive sciences (including natural history) as against *les sciences mathématiques* which in addition to pure mathematics also took in mechanics, astronomy, optics, and engineering. Rozier's Journal was clearly not a specialized periodical in the sense of the astronomical and chemical periodicals that were soon to appear, but as McClellan (1979) has pointed out in a valuable discussion of the journal, it was partly specialized in that it excluded *les sciences mathématiques*: it was specialized from the eighteenth-century point of view.

Shortly after the successful establishment of Rozier's Journal there appeared in Germany two new periodicals which were also partly specialized but in a different way: they were both natural history periodicals. Natural history, which became very popular in the late eighteenth century, partly through the writings of Linnaeus and Buffon, was the study of "the three kingdoms of nature" and derived a certain unity from the descriptive approach common to the mineralogy, botany, and zoology of the time, and from the fact that these were all field sciences. It differed from *les sciences physiques* in that it excluded the laboratory sciences (though mineralogy sometimes brought in mineralogical chemistry as well as what we would now call geology). Natural history clearly represents a degree of specialization, though from the modern point of view only a limited degree; neverless it was to continue through the nineteenth century as a field for periodical publication.

The first of these new periodicals was *Der Naturforscher* (0593), a non-institutional journal similar to Rozier's in its frequency of issue, its high proportion of original material, and its openness to all. The editor was J.E.I. Walch, a prominent figure in the University of Jena and remembered today for his work in palaeontology (DSB). In the preface to the first issue he announced that the journal would be restricted to natural history and this policy was re-affirmed in his introduction to the tenth volume:

> The [natural history] enthusiast should not find it necessary to pay out ready money for essays which do not interest him. Moreover, the field of natural history is already a sufficiently important field that those who wish to work in one or another area of natural history should not find it necessary to cross its boundaries and venture into foreign areas. This is the reason I have often found it necessary to turn down useful and learned papers when they do not definitely belong to the subject of natural history. (Translated and quoted by Kronick, 1976, p. 106)

The second of these natural history periodicals was published by an institution, the Gesellschaft Naturforschender Freunde, established in Berlin in 1773; it began its periodical (0594) two years later. The Gesellschaft is especially significant in that it was a new kind of scientific society appearing in a city which already housed one of the leading academies. It appears to be the first specialized scientific society of any size and stability, even if the specialization was only partial. But it seems partial only to modern eyes: to contemporaries the new society had a distinct field of its own, and competition with the Royal Academy was out of the question. Relations between the two institutions seem to have been cordial: many of the members of the Academy joined the Gesellschaft and the latter's periodical sometimes reprinted articles from the Academy's *Mémoires*. Most of the contents of the Gesellschaft's periodical, however, were original and it became one of the largest scientific periodicals of the time, containing many papers of importance (e.g. those of M.H. Klaproth and D.L.G. Karsten on mineralogy). With many changes of name and some changes of fortune its periodical continued into the twentieth century.

The two areas of partial specialization which have been discussed --"physics", or the laboratory sciences, and natural history --

were combined in another German non-institutional periodical, the *Magazin für das Neuste aus der Physik und Naturgeschichte* (035) which began in 1781. Neither of its two editors was very outstanding and though the journal had an appreciable size and lasted until 1806 it does not seem to have been especially remarkable.

In Italy there were also new developments, though not at this stage accompanied by any moves towards specialization. The monthly *Scelta di Opuscoli Interessanti, Tradotti da Varie Lingue* (032) began in 1775 with the intention, indicated in its title, of printing translations of articles which had appeared in other countries. From the beginning, however, it attracted original articles in various fields, including some on electricity from Volta, and from 1776 it appeared with the sub-title *Col aggiunta d'opuscoli nuovi italiani*. In 1778 the title was changed to *Opuscoli Scelti, sulle Scienze e sulle Arti*. These changes seem to indicate that the journal had found a role which its editors did not anticipate and that the Italian scientists of the day had need of a new means of publication.

A more ambitious development was the establishment in 1782 of the Società dei Quaranta, later known as the Società Italiana delle Scienze, which began its *Memorie di Matematica e di Fisica* (037) in the same year. The founder was A.M. Lorgna, a mathematician and physicist of some standing in his time. The Society seems to have been akin to the academies of the day but with the important difference--especially significant in the case of Italy--that it became national in its scope, not being the creation of some local prince or oligarchy, and that its membership was--or became--much larger than that of the typical academy. Its *Memorie*, covering the full range of the natural sciences, became the leading Italian periodical of the time and included contributions from many outstanding figures.

While these new departures were taking shape in France, Germany, and Italy, the older forms of scientific institutions and periodicals were continuing to multiply. British science showed signs of beginning to emerge from its long eighteenth-century decline with the formation of societies in Manchester (039), Dublin (043), and Edinburgh (044), as well as the Linnean Society in London (0595). In the newly established United States an academy of arts and sciences was begun in Boston (041), a counterpart to the older foundation in Philadelphia (029). (Despite their names the two American societies were local rather than national institutions.) New societies or academies were also formed at Rotterdam (031), Prague (033, 040), and Brussels (034), while the older academies at Toulouse (036) and Padua (042), both with extensive pre-histories, issued new periodicals which, however, were soon to be disrupted by the general upheaval resulting from the French Revolution.

In Paris the Académie des Sciences and its sister academies all perished in the paroxysm of the Terror but in 1795, after it had subsided, were resurrected in a new, combined form as the Institut National. The First Class of the Institut corresponded to the former Académie des Sciences and its *Mémoires* (047) carried on from the point where the old Academy's *Mémoires* had ceased. (It resembled the old *Mémoires* even in being constantly some years in arrears!) The First Class also revived the *Mémoires des Savans Etrangers* (057). The famous Ecole Polytechnique, established at the same time as the Institut National, also issued a periodical (046) which was chiefly though not exclusively mathematical. It accepted papers only from members or graduates of the Ecole and so it never became large.

In 1816, after the final removal of Napoleon and the restoration of the monarchy, the pre-revolutionary academies were re-established but with the Institut retained in a kind of federal structure --an arrangement which has continued to the present day. The newly resored Academy's *Mémoires* (066) once again carried on from the point where its predecessor had ceased but, as we shall see later, its fortune in the nineteenth century was to be very different from what it had been in the eighteenth.

In 1788 there had been formed in Paris a small scientific society with the name of the Société Philomatique, originally no more than six earnest young men who wanted to enlarge their acquaintance with all aspects of the science of the time. The Society grew slowly and by 1791 it was able to begin its *Bulletin des Sciences* (045), originally only a newsletter appearing every few months. In September 1793 this modest little society was transformed by an influx of many of the leading scientists of France in consequence of the suppression of the Academy by the revolutionaries. Once the Institut was established, two years later, they no longer took any active part in the Société Philomatique but having served for a time as the central institution of French science it now had an established reputation. It did not of course rank with the First Class of the Institut or later with the restored Academy but there was ample room for it in the enlarged scientific community of post-revolutionary France. Its *Bulletin* came out monthly and featured scientific news and extracts from other periodicals rather than original articles. For this kind of scientific journalism there was now a substantial readership.

During the Napoleonic period there existed in Paris a society which lasted for only ten years, had only fifteen members, and produced a periodical (060) amounting to only three volumes. Despite these shortcomings the Société d'Arcueil was as important as it was unique. It has been the subject of a monograph by Crosland (1967) and the very uniqueness of the Society exempts us from attempting to do it justice here. It is relevant to mention, however, that one of the reasons for its formation was a desire to create a new vehicle for publication of important work and thereby to circumvent the exasperating delays associated with the official *Mémoires* of the First Class of the Institut.

These delays, which had been the *bête noire* of French scientific life for over a century, continued with the *Mémoires* of the restored Academy but in 1835 were at last overcome in a manner which proved to be extraordinarily effective. The *Histoire* section of the eighteenth-century Academy's *Histoire et Mémoires* had contained the official record of proceedings at the Academy's weekly meetings, but this section was dropped from the *Mémoires* of the First Class of the Institut and the restored Academy, and the records or minutes (*comptes rendus* or *procès-verbaux*) did not appear between 1795 and 1835. (They were published retrospectively in 1910-13: see in the Bibliography under Académie des Sciences.) In 1835 Arago and Flourens, the two secretaries of the Academy, succeeded in obtaining finance from the Government for their publication and the *Comptes Rendus* of the Academy

(098) began in that year. What was important was that the new periodical appeared *weekly*--because the Academy met weekly--and that it included notes or short papers by members *and also by non-members*, the communications from non-members having to be approved by official committees of the Academy (all the revelations about perpetual motion, squaring the circle, etc., being quietly ignored). Moreover there was a minimum of delay: in principle, and often in practice, communications approved by the Academy at its meeting each Monday were printed forthwith and included in the issue appearing on the following Saturday.

The success of the new periodical was so great that the number and length of the papers published in it had to be restricted. Members of the Academy were allowed a maximum of ten per year, of no more than four pages each, and non-members a maximum of five per year, of no more than two and a half pages each. (Gauja, 1934, p. 97 et seq.) There is an apocryphal story that these restrictions were imposed to put some limit to the outpourings of the mathematician Cauchy who published no less than 589 notes in under twenty years (DSB--article on Cauchy).

By the 1860s, and probably earlier, the *Comptes Rendus* of the Academy was the biggest of all scientific periodicals, a position it continued to hold throughout the rest of the century (see Table 8, p. 103, and Table 20, p. 107) -- the biggest, that is, in terms of the numbers of papers (at the end of the century it was not so big as the *Berichte* of the Deutsche Chemische Gesellschaft (0421) in terms of the number of pages). As a result of its success the Academy's *Mémoires* declined and by the end of the century it was only a remnant of its former majesty. Today the *Comptes Rendus* is still one of the biggest of all scientific periodicals and has changed remarkably little in its form and functioning since its beginning.

At the close of the eighteenth century science in Britain was beginning a revival which was to accelerate quite markedly in the following decades. The same kind of developments that had first appeared in France a generation earlier made themselves evident in the appearance of a new non-institutional journal, a close analogue of Rozier's, which began in 1797 with the title *A Journal of Natural Philosophy, Chemistry and the Arts* (049). (After the first ten years it also included much natural history.) It was invariably referred to as Nicholson's Journal, after its editor. In a penetrating study of the journal and its social context Lilly (1948) points out that in Britain at the end of the eighteenth century and in the early years of the nineteenth there was a substantial increase in what may be termed the popularity of research and in the number of persons who took some part in it. The learned societies of London, Edinburgh, Manchester, and Dublin "were not well adapted to facilitate or organize this broader scientific movement, or to make use of this new phenomenon of a large mass of work each element of which was in itself of no very great importance." Most of the chief advances in British science continued to appear by virtue of the big established societies with their procedures for the reading and publication of substantial papers "but the popular movement, with its host of small significant contributions totalling a considerable portion of the whole, had to find other organizational forms. This it did in several ways, but chiefly by the foundation of smaller scientific societies ... and through the creation of a new type of scientific journal."

Nicholson's Journal was the first of this new type and was followed a year later by the second, the *Philosophical Magazine* (050). In the twentieth century, when "*Phil. Mag.*" has become entirely devoted to physics and is one of the most celebrated journals in the whole scientific world, it is hard to realise that it began as a humble semi-popular journal for science in general; indeed in its early years it was considerably inferior to Nicholson's Journal. Other journals of a similar nature appeared some years later: Thomas Thomson's *Annals of Philosophy* (064) in 1813, David Brewster's *Edinburgh Philosophical Journal* (071) in 1819, and the *Edinburgh Journal of Science* (083), also edited by Brewster, in 1824. To these can be added a periodical published over the name of the Royal Institution but presumably on a commercial basis. It began in 1816 and was known during most of its brief career as the *Quarterly Journal of Literature, Science and the Arts* (067). The editor was W.T. Brande, professor of chemistry at the Royal Institution, assisted by his friend J.F. Daniell and later by his famous colleague, Michael Faraday. By 1830 it had become quite large but it then ceased.

The founder of the *Philosophical Magazine* was Alexander Tilloch who continued to edit it until his death in 1825. Though hardly a scientist he was a very capable journalist. Nicholson's Journal was superior scientifically and very popular, at least among scientists occupied in research. Nevertheless it did not bring Nicholson much financial return and when Thomson's *Annals of Philosophy* appeared in 1813, signifying new competition at a high level, Nicholson sold out to Tilloch. In its turn Thomson's *Annals* succumbed in 1827, despite the vigorous part it had played, and was also absorbed in the *Philosophical Magazine*. In 1832 Brewster's *Edinburgh Journal of Science* came to the same end.

These three incorporations greatly strengthened the *Philosophical Magazine* not only financially but scientifically. Tilloch had been succeeded as editor in 1825 by Richard Taylor, one of the founders of the firm of Taylor and Francis which to this day continues to publish the *Philosophical Magazine* as well as other periodicals. (William Francis, a chemist and a one-time pupil of Liebig, succeeded Taylor as editor in 1851 and took as large a part in shaping the journal in the second half of the century as Taylor did in the first half.) Taylor was a well-educated man with experience in the printing trade and became a leader in scientific publishing in England. He knew how to make use of scientific advisers. At the time when the *Annals of Philosophy* was incorporated in the *Philosophical Magazine* the editor of the *Annals* was Richard Phillips (who had succeeded Thomson in 1821) and Taylor appointed him as a nominal co-editor, actually an adviser. Phillips was a competent chemist who later became one of the founders of the Chemical Society. Likewise, when Brewster's *Edinburgh Journal of Science* was taken over in 1832 Taylor appointed him to the editorial board, and he and Phillips remained on it for many years. Brewster must have been a valuable acquisition because, as well as being one of the most outstanding scientists in Britain at the time, he was an experienced journalist, having been the editor of several journals, scientific and otherwise, and author of numerous popular books. For all these reasons, then, by the 1830s the *Philosophical Magazine* was well on the way to its future renown.

In the first years of the nineteenth century the scientific community of the United States was still in an embryonic stage but it stood at the beginning of a long period of extraordinary growth, reflecting the enormous increases in the population of the country and the exploitation of its great wealth of natural resources. Few, if any, periodicals in the nineteenth century played such a central role in the growth and development of a national scientific community as did the *American Journal of Science* (069) which began in 1818. Its founder was Benjamin Silliman, professor of chemistry, mineralogy, and geology at Yale College, and through most of the century it was known as Silliman's Journal, the elder Silliman being succeeded in the 1840s by his son together with his son-in-law, the outstanding geologist J.D. Dana. Not only did the journal serve as the main medium of scientific publication in the country but it also kept its readers well informed about scientific developments in Europe. In the early and middle parts of the century, like many non-institutional scientific periodicals of the time but perhaps more than most, it also kept its readers abreast of advances in the "useful arts" or, as we would now say, technology. Though it covered all the sciences, with some excursions into the humanities and social sciences, it had an emphasis on geology and related fields -- an emphasis which reflected not only the predilections of Silliman and his editorial successors down to his grandson, E.S. Dana, but the involvement of a large proportion of American scientists in the scientific exploration of the country.

Non-institutional periodicals covering all or most of the sciences continued to appear, of course, in the Continental countries during the early nineteenth century though the great majority of them were too short-lived to be included in the Catalogue. Those that are included comprise three Italian items (062, 091, 0113), two German (076, 084), and one Norwegian (080); most of these had ceased by mid-century. Financial failure was by far the most common reason for the termination of periodicals of this kind but sometimes there were political reasons, especially about the year 1848 (e.g. 0108 and 0113), and sometimes death intervened; for example, K.E.A. von Hoff (DSB) began a promising geological journal but it was forced to cease by the death of the publisher, and Thomas Garnett (DSB) began a general science journal but died after the first volume.

By the 1840s conditions had become extremely difficult for the non-institutional periodicals for science in general. Thereafter few new ones were begun--or at least few which left any significant traces --unless they were of the special types that we shall discuss later (i.e. the semi-popular journals and the news and review journals), and those that were still in existence found survival difficult unless they were as well established as the *Philosophical Magazine* and Silliman's Journal. The reasons were partly the competition from the ever-increasing number of institutional general science periodicals but chiefly the rapid expansion of specialized periodicals (both institutional and non-institutional) for the individual sciences. As we shall see later, during the first half of the nineteenth century the bulk of scientific publishing gradually shifted from the general science periodicals to the specialized periodicals. Non-institutional periodicals for science in general could not survive in the new climate unless they had some special feature. The historical period for the successors of Rozier's Journal was over.

General learned journals, which had been so prominent on the eighteenth-century scene, still played a small part in the affairs of science in the early nineteenth century but they were mainly used, as they still are today, for surveys or special articles of various kinds addressed to the public at large. French scientists who were literary men as well, such as Biot, Flourens, and Chevreul, often appeared in the *Journal des Savants* and similar periodicals, while some English and Scottish scientists took part in the cultural debates that were a feature of the outstanding literary reviews of the period. For the publication of original work, however, such periodicals were now of little or no significance. A few are included in the Catalogue (048, 068, 072, 095, 0134), the most notable of which is the one known in its earliest period as the *Bibliothèque Britannique* (048), a Genevan publication devoted to the task of interpreting the English-speaking world in all its aspects, including science and technology, to the French-speaking world. One of its founders was the physicist M.A. Pictet, a prominent figure in Geneva at the time and well known in scientific circles in Paris and London. The journal was very successful and attained a considerable size; Pictet of course was in charge of the science and technology section which constituted a substantial fraction of the whole.

After 1815 the special concern with British affairs was reduced and the journal was re-named *Bibliothèque Universelle*. It was still international in outlook but gradually found its place as a Swiss journal. From 1836 the editor was A.A. de La Rive, then the most prominent scientist in Switzerland, and under his direction the science section branched off to become in 1846 the *Archives des Sciences Physiques et Naturelles* (0129), leaving the *Bibliothèque Universelle* to continue as a literary journal. The *Archives*, which continued successfully into the twentieth century, was ostensibly a non-institutional journal but it seems to have received much support from some of the Swiss scientific societies.

Bibliographical publications covering all or most of the natural sciences had already appeared, as has been mentioned, in the eighteenth century (021 and 022). Two such periodicals began in the 1820s but as we can see today they were doomed from birth because the initiative in this sphere was already passing to the individual sciences. Indeed the first of these periodicals, the annual reports brought out by the venerable Academy of Stockholm (074), soon split up into a number of separate series for the individual sciences. The second (081) was even more ambitious: not only was it conceived and directed by an individual, the Baron de Férussac, but it encompassed medicine, agriculture, technology, economics, history, etc., as well as all the natural sciences. Such a programme required a small army of collaborators and a large supply of capital which de Férussac's influential connections were able to mobilise for awhile, but after seven years the inevitable happened. The natural sciences were covered by two of the eight sections of the *Bulletin Universel* and as far as they, at least, were concerned the plan appears to have been executed very well. The fascicules came out monthly, the coverage was comprehensive, and the individual entries ranged from bare references to lengthy critiques, depending on the importance of the subject; some of these critiques were significant articles in their own right. The *Bulletin* was very popular among French scientists and its disappearance was keenly felt.

An important feature of scientific life in the nineteenth century (continuing into the twentieth) was the holding of national congresses at more or less regular intervals, commonly yearly. These congresses, which were open to all and were generally held in a different city each time, did much to increase the cohesiveness of the various national scientific communities as well as giving them an opportunity to explain themselves to the general public, which was something much needed in the nineteenth century. National societies, often entitled associations for the advancement of science, were created on a permanent basis to organize the congresses and many of them have continued to the present day. The proceedings of the congresses were published as periodicals, generally annuals, which can hardly be regarded as part of the archival literature of science since their contents were largely ephemeral. However they generally contain the texts of lectures, presidential addresses, etc., which can be useful material for the historian. (The bibliography can be complicated and the material difficult to locate; there are some useful details in Sarton, 1952, pp. 112-113.)

The pioneer was the Swiss national scientific society which held the first such congress at Geneva in 1815, though publication of the proceedings as a separate periodical (086) did not begin until 1825. In Germany the idea was taken up by Lorenz Oken, editor of the important journal *Isis* which will be discussed later. Since Germany was then a collection of states rather than a nation he proposed to the Leopoldina Academy that it should act as the national society and convene congresses on the Swiss model -- an appropriate suggestion since, as we have seen, the Leopoldina had been the academy of the Holy Roman Empire, the first *Reich*. The Academy however declined and Oken went on to establish the Gesellschaft Deutscher Naturforscher und Ärzte which held its first meeting at Leipzig in 1822 and soon became one of the factors bringing forth the great age of German science. Its *Bericht* (079) was published originally in Oken's *Isis* but later separately.

The British, in turn, took the idea from the Germans and in 1831 set up a "perambulating society"--the equivalent of a *Wandergesellschaft*--with the name of the British Association for the Advancement of Science (092). For many decades it had a difficult task in trying to cultivate a sympathetic understanding of science among the general public, but from its beginning it was a mainstay of the national scientific community. The same is to be said in varying degrees for the similar bodies formed subsequently in other countries -- France (096 and 0218), Scandinavia (0107), Italy (0108), Hungary (0110), the United States (0136), Russia (0201), Holland (0254), and Australasia (0256).

We have been discussing various kinds of general science periodicals which made their appearance--or in some cases their disappearance--in the early nineteenth century. But it has to be remembered that in every period the most numerous general science periodicals were those issued by the academies and other societies. Indeed after the 1830s there were very few of other kinds established. (With the individual sciences it is quite a different matter, as will be seen in the next section.) All the academies and societies whose general science periodicals are included in the Catalogue are listed by place in Table 34 which also includes, in separate sections, universities

Table 34

General Science Periodicals Published by Institutions 1665-1900

Chronology can be inferred from the entry numbers given. Number 018 corresponds to 1750, 052 to 1800, and 0140 to 1850.

1. Academies and Societies

Arranged by place

The national societies which organized congresses are not included.

Germany

Leopoldina Academy (no fixed place) 03
Berlin 07, 017, 056, 0103
Göttingen 020, 0105, 0124
Erfurt 024
Munich 027, 063, 097
Mannheim 028
Breslau 061
Frankfurt a.M. 0104
Wiesbaden 0119
Stuttgart 0125, 0247
Leipzig (a) Academy 0126, 0139
(b) Society 0225
Giessen 0133
Wurzburg 0143, 0236
Bamberg 0150
Halle 0153
Bonn 0158
Freiburg i.B. 0162
Heidelberg 0166
Königsberg 0172
Dresden 0173
Karlsruhe 0178
Jena 0187
Erlangen 0191
Marburg 0198
Bremen 0202
Greifswald 0208
Kiel 0221
Hamburg 0226
Brunswick 0233
Frankfurt a.d. Oder 0243
Rostock 0249

Britain

London 02, 094
Manchester 039, 0167
Dublin (a) Academy 043, 0102
(b) Society 051, 0163
Edinburgh 044, 0120
Cambridge 077, 0193
Leicester 0100
Glasgow 0114
Plymouth 0177

France

Paris (a) Academy 05, 010, 018, 047, 057, 066, 088, 098. (b) Société Philomatique 045. (c) Société d'Arcueil 060
Bordeaux 09, 0160
Toulouse 036, 087
Lille 053
Nancy 054, 0222
Lyons (a) Society 058
(b) Academy 0121
Clermont-Ferrand 089
Montpellier 0131
Cherbourg 0148
Rennes 0268

Italy

Società Italiana delle Scienze (no fixed place) 037
Bologna 013, 090, 0140
Turin 025, 038, 065, 0189
Siena 026
Padua 042
Naples 070, 0115, 0174, 0175
Catania 085
Milan 0111, 0185
Venice 0112
Rome 0132, 0211, 0252
Modena 0200
Genoa 0261

United States

National Academy 0182
Philadelphia 029, 0106
Boston 041, 0127
San Francisco 0159
St. Louis, Mo. 0164
Salem, Mass. 0165
Kansas City 0204

Madison, Wis. 0209
Washington (a) Society 0223
(b) Academy 0276
New York 0230
Denver, Co. 0241
Chapel Hill, N.C. 0242
New Brighton, N.Y. 0246
Indianapolis 0264
Columbus, Ohio 0269
Des Moines, Iowa 0270
Lansing, Mich. 0272

Russia

St. Petersburg (a) Academy 012, 055, 0101 (b) Society 0214
Kiev 0210, 0265
Kazan 0215
Odessa 0219

Austria

Vienna (a) Academy 0135, 0141, 0146, 0184 (b) Assoc. 0171
Graz 0181
Innsbruck 0213

Scandinavia

Copenhagen 016, 052, 059, 082
Uppsala 011
Stockholm 014, 074, 0118, 0216
Christiania (Oslo) 0168
Helsingfors (Helsinki) 0109, 0152

Holland

Harlem 023, 0197
Rotterdam 031
Amsterdam 0155, 0266

Belgium

Brussels 034, 073, 093
Liège 0117

Switzerland

Basel 019, 099
Geneva 078
Bern 0116
Zurich 0130
Chur 0157
St. Gallen 0169

Spain

Madrid 0142

Portugal

Lisbon 0203

Baltic Countries

Danzig 075
Riga 0123
Dorpat 0151

Eastern Europe

Budapest (a) Society 0207, 0239
(b) Academy 0238, 0239
Prague (a) German Academy 033, 040, 0170 (b) Czech Academy 0262
Cracow 0199, 0224, 0259
Lwow 0227
Posen 0273

Canada

Royal Society of Canada 0237

South Africa

Capetown 0231

Australasia

Hobart 0138
Adelaide 0156
Melbourne 0161
Sydney 0195
Wellington, N.Z. 0205

2. Universities

Lund 0188
Warsaw 0212
Johns Hopkins, Baltimore 0234
Toulouse 0253
Tokyo 0232
Princeton 0258
Kharkov 0263
Kansas 0267
Jurjev (or Dorpat) 0271
California 0275

3. Various Institutions

Ecole Polytechnique 046
Royal Institution 067, 0145
Smithsonian Institution 0128, 0137, 0176
Ecole Normale Supérieure 0186
Teyler's Stichting 0196
Hamburgische Wissenschaftliche Anstalt 0245

and other institutions. As can be seen, the numbers are very large in the late nineteenth century. A great many other society periodicals, especially late nineteenth-century items, were not included in the Catalogue because they were too small. Some universities which published general science periodicals are also listed in the table; there were many other such university publications in the last decade or two of the century but most were too small to include.

Apart from the institutional periodicals the only other categories of any size that were represented in the late nineteenth century were two overlapping groups -- the first, popular or semi-popular journals addressed entirely or primarily to the general public (0122, 0147, 0149, 0179, 0190, 0194, 0217, 0229, 0235, 0257) and the second, journals of scientific news and opinion addressed primarily to the scientific community (0180, 0192, 0206, 0220, 0244, 0250, 0251, 0260). Several in the latter group have continued to the present and are among the best known periodicals in the scientific world, notably *Nature* (0206) and *Science* (0244). The difficulties in establishing non-institutional periodicals for science in general in the late nineteenth century are strikingly demonstrated by the fact that *Nature*, though it soon won the respect of the international scientific community and many non-scientists, did not show a profit for thirty years and was kept in existence only by the interest of the publisher, Macmillan, and the size and nature of his business. (MacLeod, 1969)

Conditions were difficult even for such well-established journals as the *Philosophical Magazine* and the *American Journal of Science* because by the late nineteenth century scientific publishing had moved very largely to the specialized periodicals. The *Philosophical Magazine* not only survived but proceeded serenely into the twentieth century because for a long time it had been gradually specializing, first in the physical sciences and then in physics itself. It was not for nothing that its editorial board in the middle and late nineteenth century had included such figures as John Tyndall, William Thomson (later Lord Kelvin), and G.F. Fitzgerald.

The *American Journal of Science*, on the other hand, was reluctant to specialize, perhaps because its management felt that its title committed it to being a general science journal. (In this respect the *Philosophical Magazine* was lucky. One wonders how much the word 'philosophical' in its title--put there by a journalist in 1798--had to do with it becoming a journal for physics.) As late as 1918 the editor of the *American Journal of Science*, then E.S. Dana, could say:

> The result of the whole movement [of the growth of specialized periodicals] has been of necessity to narrow, little by little, the sphere of a general scientific periodical such as the Journal has been from the beginning ... That the movement will continue ... is to be expected. At the same time it has not seemed wise, at any time in the past, to formally restrict the pages of the Journal to any single group of subjects. The future is before us and its problems will be met as they arise. At the moment, however, there still seems to be a place for a scientific monthly sufficiently broad to include original papers of important general bearing even if special in immediate subject. In this way it would seem that "Silliman's Journal" can best continue to meet the ideals of its honored founder, modified as they must be to

> meet the change of conditions which a century of scientific investigation and growth have wrought. (Dana, 1918, pp. 55-56)

The reality however was that the journal was then becoming what it is today--a geological journal, despite its name.

The effects of the movement towards specialization can be seen from Table 36 which, together with Table 35, shows that while the numbers of general science periodicals grew more or less steadily from the seventeenth century to 1900 their proportion among scientific periodicals of all types decreased from 100% in 1770 and earlier to 27% in 1900.

The category of general science periodicals has, however, not faded away even today, as can be seen from the lay-out of any periodical library. Many are still published and a small proportion of them continues to be of first importance; indeed they include some of the biggest of all scientific periodicals. Those which were foremost at the beginning of the twentieth century are shown at the top of Table 20 (p. 107). The first eighteen, down to size 9, include six academy publications, six journals of the news and review type led by *Science* and *Nature*, three annual congress series--the German, British, and French--and the three grand survivors of the non-institutional journals --the *Philosophical Magazine*, the *American Journal of Science* and the Swiss *Archives des Sciences Physiques et Naturelles*. Of the academy publications the *Comptes Rendus* of the Paris Academy is in a class of its own. Grouped together are the *Proceedings* of the Royal Society (which had replaced the *Philosophical Transactions* as the Society's main periodical; the latter however is not far down the list) and the publications of the Berlin, Vienna, and Amsterdam Academies, and the Accademia dei Lincei. The latter, after a long and chequered history, had now become the national academy of Italy. The Academies of Vienna and Amsterdam appeared on the scene relatively late (in 1847 and 1853 respectively) but they were both big academies with much of the scientific output of their respective countries concentrated in their periodicals.

At the bottom of Table 20 are a large number of very small periodicals, most of them the organs of small, local, general science societies. Their scientific value was generally slight but their significance to the historian is as indexes of the extent to which science was cultivated in particular places at particular times. Tables 34 and 35 together illustrate how science grew up in the metropolitan centres of western and central Europe, at the same time expanding to the provincial centres, and then spread further afield, on the one hand to Russia and eastern Europe and on the other to the United States and other regions of the New World.

Table 35

General Science Periodicals in Existence at Successive Times

(Counting only those included in the Catalogue)

	German	British	French	Italian	American	Russian	Other	Total
1670	1	1	1					3
1680	1	1	1					3
1690	2	1	1					4
1700	2	1	2					5
1710	2	1	3					6
1720	3	1	4	1				9
1730	3	1	5	1		1	1	12
1740	4	1	5	2		1	2	15
1750	4	1	4	1		1	3	14
1760	7	1	6	2		1	5	22
1770	9	1	6	3	1	1	5	26
1780	8	1	6	3	1	1	7	27
1790	9	4	3	6	2	1	6	31
1800	7	7	5	6	2	1	6	34
1810	7	7	11	6	2	1	8	42
1820	7	8	9	9	3	1	8	45
1830	10	9	12	10	3	1	13	58
1840	11	11	13	12	4	2	16	69
1850	17	13	15	14	9	2	24	94
1860	25	16	18	14	12	3	36	124
1870	35	21	21	17	15	4	47	160
1880	39	21	21	16	20	9	55	181
1890	47	22	23	18	25	9	61	205
1900	47	22	25	19	31	12	62	218

Table 36

Proportion of General Science Periodicals at Successive Times

	Scientific Periodicals of all Types	General Science Periodicals	
		No.	%
1770	26	26	100
1780	30	27	90
1790	41	31	76
1800	51	34	67
1810	64	42	66
1820	72	45	63
1830	112	58	52
1840	160	69	43
1850	222	94	42
1860	304	124	41
1870	394	160	41
1880	505	181	36
1890	642	205	32
1900	793	218	27

3.2 The Individual Sciences

As mentioned earlier, the specialization of scientific periodicals began in the 1770s and at first was only partial: Rozier's Journal was confined to "physics", which meant the non-mathematical sciences, and almost at the same time the first natural history periodicals (or rather the first viable ones) appeared. A further degree of specialization followed very soon afterwards with the appearance of the first periodicals for the individual sciences, represented initially by astronomy, chemistry, and geology (see Table 5, p. 99). Other sciences soon followed and in the early nineteenth century each of the main individual sciences became a separate field for periodical publication.

Before going on to discuss each of the chief sciences in turn some general observations may be made here. Just as the history of the general science periodicals is a reflection of the social history of the scientific movement -- its institutions, the kinds of support it did or did not receive from society, its waxing and waning in different countries at different times -- so it is with the individual sciences. Here comparisons can be made and they can be instructive because the individual sciences differed greatly from each other in the kinds and degree of support they received from society. For example, the natural history sciences were very popular throughout the nineteenth century and their extensive amateur fringe provided them with a great deal of social support (not always welcome to the professionals). The laboratory sciences--chemistry, physics, and physiology--on the other hand had no such amateur fringe but chemistry was originally backed by pharmacy (a larger profession, relatively, at the beginning of the nineteenth century than it is today) and soon afterwards by its applications to agriculture and then by the burgeoning industrialization. Later in the century physics too was to benefit from the development of industry, especially the electrical industry which it had created. Physiology of course was intimately related to medicine but it and the other laboratory sciences benefited chiefly from the growth of research in educational institutions, above all the German universities. Geology in its early period drew much support from mining and metallurgy, and subsequently from governmental geological surveys. Astronomy was different again: for centuries it has been patronised by governments not only because of its usefulness for navigation and cartography but also because its cosmic appeal could be utilized for the sake of political prestige. And mathematics has always derived a great deal of support from its indispensable position in education.

Many of these factors are apparent in the periodical literature and are summarised in the most general way in the ratio of institutional to non-institutional periodicals, expressed in percentage terms in Table 37. (In this Table and also Table 39 the numerical data reflect chiefly the state of affairs in the late nineteenth century because the number of periodicals was much greater then.) The sciences with the greatest degree of direct social support--in other words the most popular sciences--had the greatest number of societies and so the greatest number of institutional periodicals. Much the same picture appears from the comparison of the proportion of small periodicals in

Table 37

Proportion of Institutional Periodicals

	Total No. of Periodicals	Institutional Periodicals No.	%
Natural History	82	69	84
Geology	86	64	74
Zoology	116	85	73
Geography	32	23	72
Astronomy	25	15	60
Botany	72	35	49
Experimental Biology	55	24	44
Mathematics	49	17	35
Physics	29	9	31
Chemistry	62	19	31

Table 38

Proportion of Small Periodicals in 1901-05

Taking 'small' to mean sizes 1 and 2 in Tables 21 to 33 (pp. 108-114)

	Total No. of Periodicals*	Small Periodicals No.	%
Natural History	65	47	72
Geography	28	20	71
Geology	77	54	70
Mathematics#	41	24	59
Zoology	103	58	56
Astronomy	22	12	55
Botany	53	21	40
Physics	22	7	32
Experimental Biology	40	11	28
Chemistry	39	4	10

*The totals in this table are smaller than those in Table 37 because these figures are for the number of periodicals existing in 1901-05 whereas the figures in Table 37 are for all the periodicals included in the Catalogue, many of which ceased before 1900.

#The different position of mathematics, compared with Table 37, is due to the fact that mathematics periodicals in general are smaller--in terms of the numbers of papers though not necessarily their length--than those for the other sciences.

Table 39

Percentages of the Periodicals for Each Science Published in the Chief Countries

(The numbers of periodicals are the same as in Table 37)

	Germany	Britain	France*	Italy	America
Mathematics	20	8	14	14	8
Astronomy	24	24	8	8	24
Physics	41	14	24	7	3
Chemistry	44	13	13	5	8
Geology	16	17	9	8	10
Geography	38	9	6	6	9
Natural History	20	26	18	6	8
Botany	33	21	11	6	11
Zoology	27	16	14	4	16
Exper. Biology	45	7	18	5	7

*This comparison does not do full justice to France because so much French work in all the sciences was published in the *Comptes Rendus* of the Académie des Sciences.

Table 38 because the numerous societies for the more popular sciences were mostly small societies with small periodicals. Conversely, the sciences with the largest periodicals--taking 'large' to mean size 12 and above in Tables 21 to 33--were chemistry with five, experimental biology with four, physics with two, and astronomy and microbiology each with one.

A few generalizations may be hazarded about national differences. Table 39 sets out the percentages of the periodicals published for each science in the chief countries. The numbers are mostly rather small and so the statistical significance of the percentages is not high, but some deductions seem to be warranted. In the case of Germany there is quite a marked difference between the laboratory sciences (chemistry, physics, and experimental biology) on the one hand, and the field sciences (chiefly natural history and geology) on the other. This may be largely due to the support given to the laboratory sciences in the German universities. In the other countries the same difference does not appear, indeed in the case of Britain there is to some extent the opposite tendency. France and Italy exhibit their traditional strength in mathematics, while the emphasis on astronomy in the United States is well known (astronomy appealed to millionaires as well as to emperors and kings). It was also favoured in Britain, perhaps because of the numbers of wealthy amateurs.

National societies for the individual sciences were formed in the leading countries (and in many other countries as well) during the nineteenth century, but in Britain they were mostly formed remarkably early. The Geological Society (1807), the Astronomical Society (1820), the Zoological Society (1826), the Chemical Society (1841), and the London Mathematical Society (1865) were all the earliest national societies in their respective fields, in some cases by many years, and the Linnean Society, though preceded by a German society for natural history, was founded very early (1788). A reason for this phenomenon may be that--in marked contrast to Germany--there were relatively few non-institutional periodicals in Britain, especially in the early part of the century, and their absence left a vacuum which the societies with their periodicals were created to fill. A contributory reason may be the very high degree of concentration of cultural activity in the national capital (though the same is also true of France).

A general feature of the individual sciences that is worth noting here is the number of international congresses that were held in the second half of the nineteenth century. In most fields they took place more or less regularly every few years and their proceedings were published as periodicals. (There were however some important international congresses that were held once only, notably those dealing with problems of nomenclature in chemistry, botany, etc.) They were quite different from the national congresses mentioned earlier: the latter covered all the natural sciences, and sometimes medicine as well, and had public relations as one of their main concerns; the international congresses on the other hand were for individual sciences and issues internal to them.

The earth sciences led the way -- geodesy in 1864 (0572), meteorology in 1872 (0581), geography in the same year (0557), and geology in 1878 (0502). A series of congresses for botany was also begun early (in 1864--0714) with another for plant breeding and hybridization at the end of the century (0752) which is significant in view of the developments then taking place in genetics. In the 1880s and 1890s congresses were begun for zoology (0789 and 0844), physiology (0908), chemistry (0442), and mathematics (0324). Though astronomy does not appear in the list it had, from the nature of the subject, been a leader in international collaboration for a long time, and in the last decades of the century it gave rise to much international activity, especially in the construction of star charts.

A related development, but one which did not proceed very far, was the publication of international periodicals. There was one for the teaching of mathematics (0325), three for zoological specialties (entomology, 0828 and 0829; ornithology, 0845), and one for experimental biology (0900). International co-operation was also involved during the last years of the century in the publication by the Concilium Bibliographicum in Zurich of bibliographical serials for zoology (0802), palaeontology (0533), and experimental biology (0915, 0918, 0919). The grand climax of the moves towards internationalism in the scientific world of the late nineteenth century was the establishment of the comprehensive periodical index, the *International Catalogue of Scientific Literature*, described earlier (p. 101). Sadly, this great undertaking, and internationalism generally, came to an end in 1914.

Finally some general points should be made about bibliographical periodicals in the individual sciences. This term is used in the present work to designate what are sometimes called secondary periodicals, i.e. those which catalogue or review the original papers appearing in the primary periodicals. Such periodicals, covering science in general, had first appeared in the eighteenth century (021 and 022) but from the early nineteenth century they were confined to the individual sciences or even sub-fields thereof. A distinction can be made between abstracting periodicals, giving short summaries of the original papers, and review periodicals containing discursive accounts of developments in a particular field over a specified period of time, but the distinction cannot always be drawn because some bibliographical periodicals, such as de Férussac's *Bulletin Universel* (p. 126), contained both abstracts and reviews with all their intermediate forms.

Even the distinction between bibliographical or secondary periodicals and primary periodicals is not always clear because from the time of the *Journal des Sçavans* onward many periodicals contained mixtures of original articles and secondary material including abstracts and short reviews as well as book reviews, not to mention reprints or translations of papers which had already appeared elsewhere and sundry news items. After the early nineteenth century such mixtures are to be found mostly in minor periodicals, but many major periodicals had a separate section for abstracts. Several periodicals which in the late nineteenth century or the twentieth became important abstracting periodicals, such as *Chemisches Centralblatt* (0408), began with two sections, one for original papers and one for abstracts, and gradually reduced the size of the former section in favour of the latter. There were also cases of important journals which had a section for annual reviews as well as a section for original papers, a notable example being the *Archiv für Naturgeschichte* (0759).

Insofar as one can generalize, the predominance in the early and middle nineteenth century seems to have been with reviews of the *Jahresbericht* type, Berzelius' *Jahres-Bericht über die Fortschritte der Chemie* (0401) being the most famous example, but gradually shifting during the latter part of the century to an emphasis on abstracting periodicals. No doubt this change took place because the increasing size of the periodical literature gradually made the provision of adequate reviews more difficult while at the same time the need for abstracts, and accompanying indexes, was becoming more urgent. In the early years of the twentieth century some long-established review periodicals ceased publication (e.g. 0416 and 0877) evidently because they could no longer cope with the flood of publications in their fields. And the steadily increasing importance of abstracting periodicals in the twentieth century is well known.

MATHEMATICS

In the early decades of the nineteenth century France was the undoubted leader in mathematics, as it was in much else. The *Journal de l'Ecole Polytechnique* (046) was in practice if not in principle a mathematics periodical, and the famous School which produced it has

Table 40

Mathematics Periodicals in Existence at Successive Times

	German	British	French*	Italian	American	Russian	Other	Total
1810			1					1
1820			1					1
1830	1		1				1	3
1840	1	1	1				–	3
1850	2	1	2	1			–	6
1860	3	1	2	1			1	8
1870	5	3	2	3		1	2	16
1880	8	3	4	4	2	2	5	28
1890	8	4	4	4	2	2	8	32
1900	10	4	6	6	4	2	13	45

*Not counting the *Journal de l'Ecole Polytechnique* or the *Annales de l'Ecole Normale Supérieure*.

continued to be a nursery of mathematicians to the present day. The first non-institutional journal for the subject (0277) was also French, and the second Belgian (0278), but neither survived very long, perhaps because they were before their time; however Liouville's Journal (0280) and the *Nouvelles Annales* (0283) both flourished. After 1835 a large amount of French mathematics was also published in the *Comptes Rendus* of the Académie des Sciences.

From the 1830s German mathematics, then dominated by the massive figure of Gauss, was becoming of major importance and the first German journals (0279 and 0282) appeared. About the same time British mathematics, which had remained isolated from the Continental mainstream since the beginning of the eighteenth century due partly to a sterile adherence to Newtonian methods, underwent a revival which led to the first British journal (0281). Italian mathematicians, like Italian scientists generally, were much distracted by the political upheavals which were convulsing their country, but the first Italian journal (0284) was the harbinger of an important school of mathematics in Italy later in the century.

Institutional periodicals, apart from the special case of the *Journal de l'Ecole Polytechnique* and the similar *Annales de l'Ecole Normale Supérieure* (0186), began to appear in the 1860s. Nearly all of them were publications of mathematical societies located in major centres -- London (0289), Moscow (0290), Paris (0297), Prague (0298), Hamburg (0299), Kharkov (0306), Edinburgh (0308), Palermo (0310), Tokyo (0311), Halle (0313), and New York (0317). Some of these were explicitly national societies while others were in fact if not in name (notably 0289 and 0310). In the case of Hungary the national academy was the issuing body (0318).

The non-institutional periodicals begun in the second half of the century also exhibit a considerable geographical spread -- German (0285 and 0294), British (0287), French (0315), Italian (0288, 0316, 0322), American (0301, 0305, 0320), Swedish (0292 and 0307), Danish (0286), Dutch (0302), Belgian (0303), Austrian (0314), and Polish (0312 and 0323). In general the smaller and less advanced countries were better represented in the case of mathematics than in the other sciences. This may have been a consequence of the universal importance of mathematics in education.

There were three bibliographical periodicals (0293, 0296, 0319) at least two of which had some degree of institutional support. There were however no periodicals of the annual review type such as appeared in most of the other sciences, doubtless because the character of mathematics is such that annual reviews of its progress are hardly feasible. Reviews of aspects of the subject were published as monographs rather than in periodicals. In general, books are more important in mathematics than they are in other sciences (Brown, 1956, pp. 77-78). This may be related to the fact that mathematics periodicals, when measured in terms of the number of papers, tend to be smaller than those for the other sciences (see Table 21, p. 108, in comparison with the following tables).

Two unusual features of the literature of mathematics in the nineteenth century are periodicals for the teaching of the subject (0295, 0300, 0321, 0325), obviously addressed to the relatively large numbers of mathematics teachers in secondary and tertiary education, and for the history of the subject and its applications (0291, 0304, 0309). It has, of course, a great history and nineteenth-century studies of the history of science were dominated by the history of mathematics and astronomy.

ASTRONOMY

Periodicals of the usual scientific type began in astronomy in 1800 but they had a pre-history of a unique kind. The oldest of the sciences, astronomy was sufficiently developed by the seventeenth century to be of practical value, especially for navigation and cartography. For these purposes there was a need for tables of positions and times of rising and setting of the sun, moon, and other heavenly bodies. Ephemerides containing this and related information were published annually, the first of them being the *Connoissance des Temps* (0351) which began in 1679 and in 1702 came under the control of the Académie des Sciences, thus becoming an official publication. Others followed in various countries, some with an official or semi-official status, notably the *Nautical Almanac* (0354) and the Berlin Academy's *Astronomisches Jahrbuch* (0356), not to mention later examples.

Though they were periodicals these ephemerides contained--at least in principle--no new knowledge, their contents being numerical data of a standardised kind (though the data were produced by elaborate and sophisticated calculations based on the most accurate astronomical observations available). In the 1760s and 1770s, however, some of them (0351, 0353, 0355, 0356) began to include items of astronomical news together with reports of new observations and memoirs, either reprinted from the general science periodicals or original. This was especially the case with the *Astronomisches Jahrbuch* which proclaimed its dual role in its title. It continued in its two roles from its beginning in 1774 until 1826; thereafter a new series with a new editor began and it ceased to be anything but an ephemeris. The *Connaissance des Temps* however continued to include astronomical papers until the late nineteenth century.

These two ephemerides, then, together with the minor ones, can be regarded as, in a sense, the first astronomical periodicals and the first of all the specialized periodicals (cf. Table 5, p. 99). They already existed for an astronomical purpose and it is not surprising that as the number of astronomers increased in the late eighteenth century the ephemerides began to acquire an additional, different function. After the 1820s however this function was taken over almost entirely by periodicals of the normal scientific type.

The first of these were two short-lived items (0326 and 0327), one succeeding the other and both edited by the pioneering F.X. von Zach. A much more solid beginning resulted from the establishment in 1820 of the Astronomical (later Royal Astronomical) Society of London, one of the earliest societies for an individual science. It began two periodicals, its *Memoirs* (0328) and *Monthly Notices* (0330), the latter being originally intended merely to provide brief accounts of the proceedings of the Society, with abstracts of the papers read at its meetings. As time went on, however, short papers began to appear in the *Monthly Notices* and after mid-century it gradually replaced the *Memoirs* as the chief publication of the Society. This was due to the frequency and regularity of its publication, and for the same reason it often included papers by foreign astronomers. It has continued to the present day as one of the world's leading astronomical periodicals.

The 1820s also saw the establishment of a non-institutional journal, *Astronomische Nachrichten* (0329), which immediately became the mainstay of German astronomy, a position it continued to hold through the century and beyond. (After the Astronomische Gesellschaft was founded in 1863 it took over the ownership of the journal.) It appeared weekly and as well as short papers, notes, and reports of observations, it included news items, personalia, book reviews, etc. It was international in outlook and included many contributions from foreign astronomers.

In the United States, which towards the end of the century was to become a leader in astronomy, the first periodical for the subject was Gould's *Astronomical Journal* (0331), closely modelled on the *Astronomische Nachrichten*. It began in 1849 but ceased in 1861 as a result of the Civil War and was not resumed until 1887. Despite this wide gap it did much to stimulate the development of the subject in the country, especially in its early period. It is today a leading journal. In the late nineteenth century another major journal to appear in the United States was the one which came to be known as the *Astrophysical Journal* (0339).

Publications of observatories constitute a distinct category of periodicals peculiar to astronomy. At the beginning of the nineteenth century there were about twenty-five observatories in existence and by the end of the century the number had increased to two hundred

Table 41

Astronomy Periodicals in Existence at Successive Times

	German	British	French*	Italian	American	Russian	Other	Total
1800	1							1
1810	1							1
1820	(1)							1
1830	1	2						3
1840	1	2						3
1850	1	2			1			4
1860	1	2		1	2			6
1870	3	2		1	1			7
1880	3	3		2	1			9
1890	3	3	2	2	5		1	16
1900	5	6	2	2	6	1	1	23

*Some French work was published from the late eighteenth century onward in the *Connaissance des Temps* which, with the other ephemerides is not counted here. After 1835 a substantial amount appeared in the *Comptes Rendus* of the Académie des Sciences.

(Herrmann, 1973), not counting the many private observatories. Many of them issued periodicals but these were mostly administrative reports or records of routine observations, and only a few (0332, 0333, 0340, 0342) are included in the Catalogue.

The Royal Astronomical Society remained the only one of its kind for a long time but in 1863 it was joined by the German national society, the Astronomische Gesellschaft. In its own name the Gesellschaft published only a *Vierteljahrsschrift* (0334) as it had adopted the *Astronomische Nachrichten* as its principal organ. A few years later the Società degli Spettroscopisti Italiani (0336) was established and became in effect the Italian national society. Its name is a witness to the impact that spectrum analysis had made on astronomy. In 1887 the Société Astronomique de France (0341) was founded, followed soon afterwards by the first American society (0343) and the Russian national society (0347).

By the late nineteenth century astronomy, like the other sciences, had become quite professionalized, but a characteristic of the subject was--and still is--the number of amateurs attracted to it, many of them very competent. The societies and periodicals mentioned above were of course the territory of the professionals but in the 1890s societies were established for amateurs in London (0345) and Berlin (0346). There were also some provincial amateur societies, the one at Leeds (0349) being the most substantial. The *Journal* of the British Astronomical Association (0345) continues today as one of the few astronomical periodicals that publishes results from non-professionals. The subject also appealed to a much wider public and besides the many popularizations in book form there were also periodicals (0335, 0338, 0350). There was also much astronomy in the popular periodicals for science in general (e.g. 0194 and 0235).

In Britain there was from 1877 a monthly journal of the news and review type, the *Observatory* (0337), and in 1890 the subject acquired a *Jahrbuch* (0344).

PHYSICS

In the history of science before the middle of the nineteenth century the word 'physics' can be very misleading. It is instructive to remember the description that the Royal Physical Society of Edinburgh (0766) gave of itself--"For the promotion of zoology and other branches of natural history." Here the word 'physical' means 'scientific', or more exactly 'scientific but non-mathematical'. Confusion is compounded by the fact that the older and newer meanings of the word overlapped. For example, the word is used in the modern sense in the name of the Physikalische Gesellschaft zu Berlin, founded in 1843, while it is used in the old sense in the name of the Physikalisch-Medicinische Gesellschaft in Würzburg, founded in 1849. Indeed something of the old meaning can remain today in the term 'the physical sciences' which is understood by different people in different ways. Augmenting the confusion still further is the fact that, though physics was the "glamour science" of the mid-twentieth century and in the eyes of some people (especially philosophers of science) is the only science worth the name, it was at the beginning of the nineteenth century "an immature, undisciplined pursuit with indefinite limits and little cohesiveness among its various concerns" (Silliman, 1974, p. 138). In so far as it was a field at all it was then a minor one compared with chemistry which was the "glamour science" of the time.

It is not surprising, then, that there were very few periodicals for physics in the first half of the century, and indeed few before the 1880s (Table 42). There was one however which began as early as 1790 and became quite outstanding in the early nineteenth century as well as later. It originated as the *Journal der Physik* (0362) and when its founder, F.A.C. Gren, died in 1798 it was taken over by L.W. Gilbert who called it the *Annalen der Physik*, by which title it has since been known (though, as we shall see, there was later some tampering with the title). Gilbert was succeeded as editor in 1824 by J.C. Poggendorff who edited it for fifty years; he was followed in 1877 by G. Wiedemann who continued as editor to the end of the century. In the early twentieth century it was one of the most famous of all scientific periodicals, especially after it had published Einstein's theory of relativity.

In its earliest years, under Gren's editorship, its scope was very wide though the majority of the contents were what we would call physics. There was much chemistry, however, as there continued to be for several decades. Gilbert in 1819 extended the journal's title to *Annalen der Physik und physikalischen Chemie*; the latter phrase

Table 42

Physics Periodicals in Existence at Successive Times

	German	British	French	Italian	American	Russian	Other	Total
1790	1							1
1800	1							1
1810	1							1
1820	1							1
1830	1						1	2
1840	2						1	3
1850	2			1			–	3
1860	2		1	1			–	4
1870	3		1	1			–	5
1880	3	3	4	1			–	11
1890	8	3	5	2		2	1	21
1900	8	4	7	2	1	2	1	25

did not mean physical chemistry in the modern sense but rather the area that was then regarded as belonging to both physics and chemistry. When Poggendorff took over he changed the title to *Annalen der Physik und Chemie* and the words "und Chemie" remained until 1900 when a new series began and they were dropped. The inclusion of chemistry was presumably due to the fact that so many physicists then were also chemists, and vice versa (Faraday provides an outstanding example).

We find a similar situation reached from the opposite direction in the case of the important *Annales de Chimie* (0393) which from its beginning in 1789 until 1815 was very largely, if not entirely, a chemistry journal. In the latter year however the board of editors decided, in a re-organization involving several changes, to alter the title to *Annales de Chimie et de Physique*. This title it retained until 1914 when it divided at last into two journals, one for chemistry and one for physics. Another example of a chemistry journal co-opting physics is provided by the *Journal für Chemie und Physik* (0397). One possible reason why journals increased their scope in this way is of course the expectation that they would thereby gain additional readers and hence additional subscriptions. In the case of the *Annales de Chimie*, however, there seem to have been reasons of principle rather than expediency -- a conviction that it would be better for the advancement of knowledge if physics and chemistry were studied together (Court, 1972, pp. 127-128).

With the *Annalen der Physik* physics in Germany was well catered for, and all the more so when the Physikalische Gesellschaft zu Berlin was established in 1843. Because of the existence of the *Annalen*, with which it became closely associated in 1877 (if not before), the Gesellschaft did not need to establish a periodical of its own of the usual type; instead it instituted an authoritative annual survey of the literature of the field (0366). Dove's *Repertorium* (0364) had been to a large extent a one-man effort in this direction and was superseded by the new review.

In Italy in 1844 a journal (0365) with the evocative name of *Il Cimento* ("Experiment") was established by the important physicist and physiologist, C. Matteucci. It was short-lived, possibly because of Matteucci's involvement in the political upheaval of 1848, but in 1855 he revived it with the title *Il Nuovo Cimento* and it soon became an imortant journal. Though it covered both physics and chemistry the former predominated and by the end of the century its contents were entirely physics. In the early twentieth century it became the organ of the Società Italiana di Fisica.

The *Journal de Physique Théorique et Appliquée* (0370), which became the chief French journal for physics, was begun in 1872 by J.C. d'Almeida. In the following year the Société Française de Physique (0371) was established, d'Almeida being one of the founders; in the twentieth century it took over the *Journal de Physique*. At almost the same time the Physical Society of London was formed; its *Proceedings* (0372) were small, doubtless because of the existence of the *Philosophical Magazine* which contained much physics as did the *Proceedings* of the Royal Society. By the end of the century German physics had become so big that it was able to support another large periodical, the *Physikalische Zeitschrift* (0390), in addition to the *Annalen der Physik* (0362 and 0373) and the *Verhandlungen* of the Physikalische Gesellschaft (0381).

In Russia physics was the junior partner of chemistry: the Russian Chemical Society, established in 1869, expanded to become the Physical-Chemical Society (0422) in 1873 but by the end of the century the amount of physics in its periodical was only a quarter of the amount of chemistry. Nor was physics very strong in the United States in the nineteenth century, compared with astronomy and some of the other sciences. The only American physics periodical of any significance was the *Physical Review* (0388) which began in 1893; it later had the support of the American Physical Society which was established in 1899. Even by 1940 American physics was not very impressive and no one could have foreseen the enormous growth that was to take place after the Second World War. A consequence of this growth is that the *Physical Review* is now the world's leading periodical in the subject.

Some degree of specialization began to appear in the physics periodicals in the latter part of the century, as is indicated by the commencement in 1880 of the *Central-Zeitung für Optik und Mechanik* (0376). It seems to have been largely concerned with instrumentation and an important periodical for this field, the *Zeitschrift für Instrumentenkunde* (0380), began in the following year; an earlier periodical in the field (0368) ceased in 1891 presumably because of the new competition. Specialization also appears in the rather unusual case of the *Communications from the Physical Laboratory at the University of Leiden* (0383). A German journal for the teaching of physics and chemistry (0385) began in 1887 and there was a contemporary Russian journal (0384) which also seems to have been addressed to teachers or to the general public.

One of the main reasons why physics was so much smaller than chemistry during most of the nineteenth century was that physics did not have the extensive hinterland in industry (to say nothing of agriculture and medicine) that chemistry had, nor did the applications of physics in engineering have the close connections with the pure science that was characteristic of most applications of chemistry. A relatively abrupt change in the fortunes of physics in this regard followed from the invention of the electric lamp in 1878. The resulting demand for electricity for lighting triggered an extremely rapid development of the electric power-generating industry during the 1880s and 1890s, and consequently of the technology of electrical engineering. Hitherto the chief application of electricity had been in telegraphy and we find some of the journals for telegraphy including articles of some significance for physics (0367 and 0369). Telegraphy was an important use but a very small one compared with electric lighting and later electric power for locomotion and industrial use generally. After 1878 there was a proliferation of periodicals of all kinds dealing with the new industry. Several which seem to have had fairly close relations with the parent science have been included in the Catalogue (0374, 0375, 0377, 0378, 0379, 0382, 0386, 0387) but many more were excluded.

It was chiefly in consequence of the development of the electric industry that important national laboratories for applied physics were set up around the end of the century -- in Germany the Physikalisch-Technische Reichanstalt (cf. 0380) in 1884, in Britain the National Physical Laboratory in 1899, and in the United States the Bureau of Standards in 1901. The symbiotic relationship between the new technology of electrical engineering and its parent science is illustrated by the joint action of the British Institution of Electrical Engineers and the Physical Society of London in setting up an abstracting journal covering both fields (0389b). The place of electricity in the nineteenth-century literature of physics is demonstrated by the fact that the sub-section for electricity and magnetism in the Physics section of the subject index of the Royal Society's *Catalogue of Scientific Papers* constitutes 41% of the whole section. (The Royal Society's *Catalogue* covers the entire century but the great bulk of the entries date from the last decades.)

CHEMISTRY

Through the nineteenth century and beyond chemistry was the biggest of the physical sciences, the only other science comparable with it in size being zoology. During the course of the eighteenth century it had gradually emerged as a distinct science with some measure of respectability, having outgrown its origins in craft empiricism and the dubious practices of the alchemists. It was still intimately connected with pharmacy, however, and this department of the profession of medicine provided most of its social support until the advent of industrialization, the teaching of the subject for example being nearly always carried on in the medical schools. It was also taught in the mining schools, as part of mineralogy and metallurgy, but such schools were still few in number.

Though the most important theoretical developments in the late eighteenth century came from Britain and France, chemistry seems to have been especially strong in Germany in terms of the number of its practitioners--nearly all pharmacists--and in social esteem (Hufbauer, 1971). In Robert Boyle's day Germany was "the land of the alchemists" and up to the early twentieth century, if not still, chemistry has been Germany's *Lieblingswissenschaft*. And it was in Germany that the first chemistry periodical (0391) appeared in 1778, begun by Lorenz Crell, a professor of medicine and pharmacy at one of the minor universities. If we do not count the astronomical ephemerides discussed earlier, Crell's Journal is the first periodical to be devoted to an individual science. It was soon followed by numerous other German journals dealing with chemistry (though most of them called themselves pharmacy journals) but before we look more closely at them it is necessary to consider the developments in France which were turning the theory of the subject almost literally upside down.

Lavoisier's chemical revolution had taken shape during the 1770s and 1780s and the new nomenclature, in which the new theory was embodied, was promulgated in 1787 and followed two years later by Lavoisier's classic *Traité Elémentaire de Chimie*. It took some years for the chemical world to be converted to the new view of the subject which was so different from the old one associated with the phlogiston theory, and in all the leading countries there were debates between the "antiphlogistonists" and the "phlogistonists". It was in this milieu that the *Annales de Chimie* (0393), the first--and for a long time the chief--French periodical for chemistry, was born in 1789. At that time the most important periodical in France, apart from the *Mémoires* of the Academy, was the *Journal de Physique*, the title now carried by what had been Rozier's Journal (030). The editor now was Lamétherie who was bitterly opposed to Lavoisier and the new chemistry and refused to publish any papers using it. As a result Lavoisier and his associates instituted the new journal.

In announcing their aims in the first number the editors of the *Annales de Chimie* emphasized that they would respect the opinions of their contributors who were free to use whatever chemical language they chose. In fact, however, only papers embodying or supporting the new view appeared in the journal (apart from some abstracted or reprinted from elsewhere). Chemists who adhered to the phlogiston theory sent their papers to the *Journal de Physique*. This seems to be the first case in science of competing periodicals supporting competing theories. The situation did not last long, however, for in a few years the new chemistry had won. But Lamétherie never accepted it and his journal must have suffered in consequence. He died in 1817 and the *Journal de Physique* ceased soon afterwards -- an ignominious end to a periodical which had been so innovative in its early years.

France continued to be the leader in chemistry up to the 1830s and during the period the *Annales de Chimie* was the most prestigious journal in the chemical world (after 1815 it also included physics but chemistry always predominated). The only other French chemical periodicals of any significance during the period were the *Bulletin de Pharmacie*, later the *Journal de Pharmacie* (0399), and the less important *Journal de Chimie Médicale* (0405). In Italy Brugnatelli's *Annali di Chimica* (0394) seems to have been modelled on the *Annales de Chimie* but it did not last long and the main Italian periodical

Table 43

Chemistry Periodicals in Existence at Successive Times

	German	British	French	Italian	American	Russian	Other	Total
1780	1							1
1790	2		1					3
1800	5		1	1				7
1810	4		2	-			1	7
1820	5		2	-			-	7
1830	7		3	1	1		1	13
1840	6	1	4	1	1		1	14
1850	6	4	4	1	1		1	17
1860	7	3	5	1	1		1	18
1870	9	3	5	1	1	1	-	20
1880	10	4	4	2	3	1	1	25
1890	13	6	4	2	3	1	5	34
1900	20	6	7	2	5	1	7	48

was for pharmacy (0404). A pharmacy periodical of some small significance for chemistry began in the United States in 1825 (0406) but it was not until as late as 1841 that a significant one appeared in Britain (0413). In Germany, however, during the late eighteenth and early nineteenth century the pharmacists were extraordinarily active in establishing periodicals and it was out of their proliferation and complicated transformations and amalgamations that there emerged, as we shall see, three journals of major importance for chemistry.

Following Crell's successful initiative several pharmacy journals were established around the turn of the century, the most important of them--up to the early 1830s--being Trommsdorff's (0395). J.B. Trommsdorff and F.A. Gehlen were typical of these apothecary-journalists, and like some of the others they extended their activities to more than one journal. Thus Trommsdorff edited 0392c as well as 0395, and Gehlen at various times was concerned with 0396, 0397b/c, and 0400. Crell, like all his contemporaries, was a phlogistonist but unlike many of them he remained so, and continued to oppose the new French doctrine even in the late 1790s when it had become widely accepted in Germany. This must have diminished the status of his journal, hitherto so successful, and in 1804 it succumbed to the competition of the newcomers, especially the *Allgemeines Journal der Chemie* (0397) which had been launched under the antiphlogistonist banner in 1798. This journal, renamed the *Journal für Chemie und Physik* (0397c/d) became, especially under the editorship of J.S.C. Schweigger, one of the leading German periodicals of the time and was devoted as much to physics as to chemistry.

In 1825 Justus Liebig, then only twenty-three years old, became professor of chemistry at the small university of Giessen and began the career that was soon to make him the most outstanding figure in chemistry, especially in the big new field of organic chemistry, and eventually one of the best known scientists of the nineteenth century. His origins were in the apothecary tradition: his father had been an apothecary and he himself had been an apprentice before he went on to higher things. It is not surprising, then, that in 1831 he became involved in the editing of a small pharmacy journal (0403c) which, as a result of a merger, became in 1832 the *Annalen der Pharmacie* (0409). In 1838, after a succession of editorial re-arrangements, sometimes acrimonius, Liebig was left in full control of the journal. He appointed as co-editor his friend Friedrich Wöhler, who was also in process of becoming an outstanding chemist, and changed the title to *Annalen der Chemie und Pharmacie*. (Eventually it completed the transition and became *Annalen der Chemie*.) In a few years the *Annalen* became the most important of all chemistry periodicals of the time, partly because of Liebig's editing, partly because of the fame of his researches and of the remarkable school that he had built up at Giessen, and partly because the leadership in chemistry (as in science generally) had now passed from France to Germany.

A close second to the *Annalen* was the *Journal für praktische Chemie* (0407b), also formed by a merger in the early 1830s and also still in existence today. It was edited by O.L. Erdmann who was professor of chemistry at Leipzig and founder of a research school which was able to compete with Liebig's. Another journal, important in a different way, which was founded at about the same time was the *Pharmaceutisches Central-Blatt* (0408). As with Liebig's *Annalen*, the changes in its title graphically illustrate chemistry's detachment from pharmacy: in 1850 it became the *Chemisch-pharmaceutisches Central-Blatt* and in 1856 the *Chemisches Central-Blatt*. It included abstracts as well as original papers and the size and importance of its abstracts section increased through the century. In the late nineteenth and early twentieth century it was probably the biggest abstracting journal in science (as its American counterpart, *Chemical Abstracts*, now is).

Another important bibliographical development was the publication of annual reviews of progress in chemistry. We have already noticed that the Stockholm Academy in 1821 began an annual series of reports on the progress of science (074) which was necessarily arranged under the headings of the individual sciences. In 1818 the great Swedish chemist, Berzelius, had become secretary of the Academy and it may have been his influence which led it to undertake the project. In any case Berzelius wrote the reports on chemistry (0401) which, with those on botany (0686), were the chief outcome of the project, both being translated into German. It was a very valuable initiative and the German version of Berzelius' reports was widely read and had a powerful influence. (In the 1840s there was also a French version.) Berzelius had to give up the task in 1846, shortly before his death, and Liebig and Kopp established a German annual (0416) to carry it on as a collective effort. For many years Kopp acted as general editor, assisted by numerous collaborators.

In Britain, where the apothecary tradition had never been strong, at least journalistically, no chemical periodical of any significance appeared before 1840. There was plenty of chemical research going on but the results were published in the general science periodicals such as the *Philosophical Magazine* and its early rivals. Perhaps it was the absence of a strong non-institutional chemistry journal like the *Annales de Chimie* or Liebig's *Annalen* which led to Britain being the first country to create a national chemical society. The Chemical Society's *Memoirs* (later *Journal*) (0412) was begun in 1843 and has continued to the present as the hub of British chemistry. At almost the same time three other British periodicals appeared (0411, 0413, 0414) at least one of which was addressed largely to the world of industry. With a French journal of a similar kind appearing at the same time (0410) it is an indication of the importance to chemistry of the rapidly spreading industrialization.

The French national chemical society, the Société Chimique de Paris, was established in 1858 and its *Bulletin* (0417) gradually overtook the *Annales de Chimie*. The fortunes of German chemistry at this time were largely personified in the career of A.W. von Hofmann who presided over the emergence of coal-tar chemistry and in 1865 took up the key position of professor of chemistry at the University of Berlin, the centre of the expanding German university system, and two years later was chiefly instrumental in establishing the Deutsche Chemische Gesellschaft. The extremely rapid industrialization of Germany was then getting under way, with the chemical industry--spearheaded by the new synthetic dye industry--a prominent part of it. In chemistry as in other fields it was an enormously productive era and the *Berichte* of the Deutsche Chemische Gesellschaft (0421) doubled in size between 1870 and 1880 and by the end of the century, if not before, it ranked with the *Comptes Rendus* of the Académie des Sciences as the largest of all scientific periodicals (though smaller than the *Comptes Rendus* in terms of the number of papers it was larger in terms of the number of pages). The *Berichte* continued to be the world's most important chemical periodical until the Second World War.

In other countries in the second half of the nineteenth century national chemical societies were set up, or alternatively non-institutional journals became the unofficial or semi-official organs of the country's chemical community -- Russia (0422), Italy (0424), Bohemia (0425), United States (0429 and 0430), Austria (0431--the national academy being the issuing body), Japan (0432), Holland (0434), Sweden (0438), and Hungary (0449). In the United States both courses were taken, the non-institutional *American Chemical Journal* (0429) being initially the more important but gradually yielding the leadership to the *Journal* of the American Chemical Society (0430) which in the second half of the twentieth century has become the world's leading chemical periodical.

An important feature of the periodical literature of chemistry in the second half of the century, and a testimony to the growth and development of the subject, was the appearance of periodicals for sub-fields -- analytical chemistry (0420, 0426, 0451), physiological chemistry, now called biochemistry (0423, 0427), physical chemistry (0437, 0443, 0444, 0450), and inorganic chemistry (0440) which had been the first part of chemistry to be established on a sound theoretical foundation (chiefly by Lavoisier) but was now much smaller than organic chemistry. Such an array of specialized periodicals was surpassed only by zoology. Most of them were of high quality and some became quite big (notably 0427, 0437, and 0440). In the latter part of the century Liebig's *Annalen* became exclusively devoted to organic chemistry and this, the largest part of the subject, constituted the bulk of the contents of the general chemical journals.

In the last few decades of the century there was a multiplication of all kinds of periodicals dealing with the chemical industry and various aspects of applied chemistry. Because the relations between pure and applied chemistry are generally quite close a number of these had some relevance to pure chemistry, especially the *Journal* of the Society of Chemical Industry (0433) and the *Zeitschrift für angewandte Chemie* (0439). A notable field of applied chemistry--and one of the areas in which Liebig had been a pioneer--was food chemistry (0436 and 0447). The striking effect that the electrical industry had on physics (p. 147) is reflected in the rapid growth of electrochemistry, pure and applied (0443, 0444, 0445, 0448), the development of electrochemical industries having been made possible by the new availability of electric current in large quantities.

Bibliographical services were increasingly needed as the chemical literature continued to expand. *Chemisches Centralblatt* (0408) and the *Jahresbericht über die Fortschritte der Chemie* (0416b) became bigger and bigger and were supplemented by other review serials (0441, 0452, 0453), including some for specialized fields (0423 and 0443). Several of the main periodicals for general chemistry also included abstracts. In the United States a small review begun in 1895 (0446) was refashioned a few years later as *Chemical Abstracts*, today the colossus of the scientific literature. An important function was also performed by the news periodicals (0419 and 0428) which now had a big readership as there were so many chemists employed in industry.

GEOLOGY

The formative period of geology can be compared with that of chemistry. Geology too had emerged as a more or less distinct science by the late eighteenth century, perhaps a few decades later than chemistry, and just as chemistry had its main backing in the long-established profession of pharmacy so geology drew most of its support from the no less ancient professions of mining and metallurgy. And as pharmacy and its periodical literature had flourished especially in Germany, so it was with mining and metallurgy in which Germany had been the leader for many centuries.

Such parallels must not be pushed too far however. Geology had a local or regional emphasis quite inapplicable to chemistry, and it was much smaller. Institutions and their periodicals were to play a much bigger part in its literature. And after the early nineteenth century the German contribution was to be much less than in chemistry and the British contribution proportionately greater. Moreover since the middle of the nineteenth century geology has been unique among the natural sciences in the high proportion of governmental publications in its literature.

The first periodical of significance for geology was the *Bergmännisches Journal* (0454) which began in 1788 and emanated from the mining academy at Freiberg in Saxony which was soon to become famous as a result of the work of A.G. Werner, one of the main pioneers of geology and a highly regarded and influential teacher. The journal was founded and edited by A.W. Köhler and C.A.S. Hofmann, two of Werner's students who were closely associated with him. It ceased in 1815 but in 1827 the Freiberg Academy instituted an annual review of mining and metallurgy (0463) which continued into the twentieth century.

Among the scientific and technical institutions set up in Paris in 1794, in the constructive phase of the Revolution (institutions which included the Institut National, the Ecole Polytechnique, the Bureau des Longitudes, etc.) there was, besides the Ecole des Mines, an Agence (later Conseil) des Mines which began a periodical with the title of *Journal des Mines* (0455), changed some twenty years later to *Annales des Mines*. Besides mining engineering it included much mineralogy and geology: for a long time some of the most eminent mineralogists and geologists in France were members of its editorial board, and for several decades it ranked as one of the most important periodicals for the geological sciences. Other early mining periodicals included a Swedish one (0458), a Russian one (0461) on the French model, and a German one (0460) edited by the leading metallurgist, C.J.B. Karsten. A few of some significance for geology were established later in the nineteenth century (0471, 0476, 0477, 0478, 0479).

The first purely mineralogical and geological periodical was the *Taschenbuch für die gesammte Mineralogie* (0456) founded in 1807 by the important geologist K.C. von Leonhard and edited by him until his death in 1862. The term 'mineralogy' then covered what we would now call geology and palaeontology, as well as mineralogy in the modern sense of the word. In some quarters, especially in Germany, the term 'geology' was rather suspect because it was associated with the highly speculative theories of the early cosmogonists, and the term 'geognosy' was used. It was presumably because of shifts in terminology that in 1830 von Leonhard changed the name of his journal to *Jahrbuch für Mineralogie, Geognosie, Geologie und Petrefaktenkunde*. From 1833 it was known as *Neues Jahrbuch....* because of some changes in its format and periodicity -- despite the title *Jahrbuch* it was now issued in bimonthly parts. After von Leonhard's death a new series was commenced by his successors in 1863 with the title *Neues Jahrbuch für Mineralogie, Geologie und Palaeontologie* as a result of further changes in accepted terminology. From its beginning the journal was very successful and it continued through the nineteenth century and beyond as one of the chief German periodicals in the field.

An important part of the literature of geology is contained in the periodicals of the various national societies. The first was the Geological Society of London (0457), founded in 1807 and one of the oldest societies for any of the individual sciences. (Its formation was viewed with disfavour by Sir Joseph Banks, President of the Royal Society, who attempted to bring it under the Royal Society's control.) It was followed in 1830 by the Société Géologique de France (0464) and in the same year a mineralogical society was established in Russia (0465); no doubt the Russian equivalent of the word 'mineralogical' in its name had the old wide meaning which covered geology and palaeontology as well as mineralogy. The Association of American Geologists (0470) began in 1840 but a few years later was enlarged to become the American Association for the Advancement of Science (0136) and it was

Table 44

Geology Periodicals in Existence at Successive Times

	German	British	French	Italian	American	Russian	Other	Total
1790	1							1
1800	1		1					2
1810	2	1	1					4
1820	2	2	1				1	6
1830	3	3	1			1	1	9
1840	3	5	3		1	2	1	15
1850	6	7	3		–	2	1	19
1860	6	10	3		–	2	7	28
1870	6	12	5		1	2	8	34
1880	7	14	6	1	1	2	15	46
1890	11	12	7	3	3	4	21	61
1900	12	13	6	7	8	6	27	79

not until 1890 that a more permanent body, the Geological Society of America (0522), was established; however American geologists always had the *American Journal of Science* (p. 125) as a publication medium. The Deutsche Geologische Gesellschaft (0474) was set up in 1848 and the second half of the century saw the establishment of national societies in Hungary (0493), Sweden (0495), Belgium (0496), Switzerland (0497 and 0517), Italy (0509), and South Africa (0532).

In Britain a remarkable number of provincial geological societies came into existence (and, as will be seen later, there was an even greater number of natural history societies). The first was the Geological Society of Cornwall (0459). Cornwall had been an important mining region since antiquity and it figured prominently in the early history of British geology. Other societies were later set up in Dublin (0467), Leeds (0468), Manchester (0469), Liverpool (0483), Glasgow (0484), and Edinburgh (0490); the Wernerian Natural History Society (0597) can be regarded as an early predecessor of the Edinburgh society. There were a few provincial geological societies in France and Germany but, with the exception of the one at Lille (0492), their periodicals were too small to be included in the Catalogue. No doubt the provincial societies were largely or wholly amateur in character. In London in 1859 the Geologists' Association (0486) was established to cater for the interests of amateurs, and a non-institutional journal (0482) at the amateur level began at the same time. Typically the amateurs were concerned with local studies. According to A.J. Meadows (1974, p. 73):

> A comparison of British geological publications over the last century and a half reveals not simply an increasing emphasis on

the primacy of research, but also a changing definition of what constitutes an acceptable research paper. Thus, in the nineteenth century, local geological studies represented worthwhile research in their own right; but, in the twentieth century, local studies have increasingly become acceptable to professionals only if they incorporate, and reflect on, the wider geological picture.

An unusual feature of the literature of geology is the very large number of publications issued by the geological surveys set up by national or state governments. The existence of these surveys is of course due to the easily perceived economic value of a thorough geological knowledge of the national territory. Britain was the first to establish a governmental survey, in the 1830s, and was soon followed by other countries and many of the American states. Often the national survey came into existence gradually, in a series of steps, and was sometimes preceded by the construction of a geological map of the country by independent geologists on their own initiative. Though these surveys were established from an economic motive the geological knowledge they accumulated could at the same time be a substantial contribution to the parent science. Much of it however was of a very local nature and of limited interest; most of the periodicals issued by geological surveys were not cited in the Royal Society's *Catalogue of Scientific Papers* or the *International Catalogue of Scientific Literature* and so were not included in Part 1. Those that were included are the following. (The dates of these entries should not be taken to indicate the dates of foundation of the surveys.) Britain (0472 and 0535), Austria (0475 and 0481), India (0480 and 0487), Italy (0491), Denmark (0500 and 0521), United States (0504--the national survey. State surveys: Indiana, 0489; Iowa, 0523; Maryland, 0536; Vermont, 0538), Prussia (0505), Holland (0507), Russia (0510 and 0511), Canada (0512), Alsace-Lorraine (0515), Bavaria (0518), France (0519), New South Wales (0520), Cape of Good Hope (0531).

Specialization in the geological literature appears chiefly in the establishment in the second half of the century of separate periodicals for mineralogy (0494, 0499, 0501, 0503, 0514, 0527) and palaeontology (0473, 0497, 0506, 0508, 0528, 0533, 0534). These however are not sub-fields of geology but related sciences (mineralogy is older than geology and palaeontology is essentially biological). At the end of the century the sub-fields of seismology (0530 and 0537) and pedology or soil science (0539) appear; the latter was for a long time a Russian speciality. There were bibliographical periodicals for palaeontology (0533) and mineralogy (0501b) as well as for geology generally (0485 and 0498).

In the last years of the century American geology was showing signs of its future leadership with the establishment of three new periodicals which were to become of major importance in the twentieth century -- the *American Geologist* (0516; in 1905 it became *Economic Geology*), the *Bulletin* of the Geological Society of America (0522), and *Journal of Geology* (0525). In addition the *American Journal of Science* (069) was on the way to becoming entirely a geological journal, and a major one.

GEOGRAPHY

We are concerned here only with physical and mathematical geography but nearly all the periodicals containing papers dealing with these fields also included a great deal of descriptive and economic geography. Most of the periodicals were institutional, chiefly those of the national societies -- France (0540), Britain (0541, 0550, 0564), Germany (0542, 0548, 0558), United States (0565), Austria (0552 and 0571), Russia (0554), Italy (0556), Holland (0559), and Denmark (0560). As can be seen from Table 45 German publications predominated. There were three German provincial societies (0543, 0549, 0561) and two special kinds of institutions (0551 and 0563). One governmental organization is included -- the United States Coast Survey (0546), a pioneering effort in governmental science.

Of the nine non-institutional periodicals, two (0566 and 0569) might be described as professional together with another on the special subject of hydrology (0570), one is a *Jahrbuch* (0555), and the others are various kinds of reviews or popularizations (0545, 0553, 0562, 0657, 0568).

Table 45

Geography Periodicals in Existence at Successive Times

	German	British	French	Italian	American	Russian	Other	Total
1830			1					1
1840	2	1	1				1	5
1850	2	1	1			1	1	6
1860	4	2	1		2	1	2	12
1870	6	2	1	1	2	1	2	15
1880	8	2	1	1	2	1	5	20
1890	9	2	1	1	3	1	5	22
1900	11	2	2	2	3	2	6	28

NATURAL HISTORY

As mentioned earlier, natural history derived some degree of unity from the descriptive approach which characterized the early stages of botany, zoology, and mineralogy (the last including parts of what would now be called geology) and from the fact that these were originally all field sciences and could still be cultivated as such by amateurs even after the professionals had taken them into the laboratories and museums. The broad area designated as natural history was popular all through the nineteenth century and, especially in the latter part of the century, had the most extensive amateur fringe of

Table 46

Natural History Periodicals in Existence at Successive Times

	German	British	French	Italian	American	Russian	Other	Total
1780	2							2
1790	2	1						3
1800	2	1						3
1810	1	2				1		4
1820	1	2			1	1		5
1830	1	3	5		2	2		13
1840	2	6	5	1	3	2	1	20
1850	7	7	6	1	5	2	4	32
1860	8	9	7	1	5	2	8	40
1870	10	10	7	1	6	2	9	45
1880	11	14	10	2	7	2	14	60
1890	11	17	12	4	7	2	15	68
1900	12	18	13	4	7	2	16	72

any part of the natural sciences. For this reason the great majority of natural history periodicals were the publications of societies, many of them rather minor provincial societies (a large proportion of which were excluded from the Catalogue because of their smallness). Natural history represents only a partial degree of specialization, intermediate between science in general and some of the individual sciences, and many of the periodicals are difficult to distinguish from general science periodicals; in the names of many of the societies there is no clear distinction from science in general.

From early in the eighteenth century there were some small natural history journals, mostly short-lived, but as mentioned earlier (p. 120) the first substantial ones were *Der Naturforscher* (0593) and the publication of the Gesellschaft Naturforschender Freunde zu Berlin (0594), both beginning in the early 1770s. The latter continued with some interruptions through the nineteenth century, and from the 1830s was accompanied by the publications of several German provincial societies (0612, 0621, 0622, 0623, 0625, 0626, 0629, 0635).

In Britain the Linnean Society was founded in London in 1788 (0595), followed by the Wernerian Natural History Society in Edinburgh in 1808 (0597), the latter being chiefly though not exclusively concerned with geology. It was in Britain that provincial societies were most numerous (0608, 0609, 0624, 0627, 0640, 0644, 0645, 0647, 0648, 0658, 0662, 0663, 0669) and, as we have seen earlier, there were also several for geology. In Paris a Société Linnéenne began in 1788, following the British example, but faded away when it was met with the hostility of the Académie des Sciences. It was revived in 1791 under the less provocative name of the Société d'Histoire Naturelle but its career continued to be troubled and its periodical (0600), after some abortive beginnings, only ran from 1823 to 1834. French provincial societies however were established from the 1820s onward (0601, 0603, 0604, 0607, 0619, 0649, 0656, 0664, 0670).

Through the century natural history societies were established in various cities of America (0598, 0599, 0610, 0642, 0655), Italy (0636, 0651, 0668), and Switzerland (0618, 0620, 0659), and at numerous other centres -- Moscow (0596), Copenhagen (0628), Batavia (0630), Prague (0631), Vienna (0633), Montreal (0638), Brünn (0641), Helsinki (0650), Sydney (0652), Melbourne (0666), and Bombay (0667). That their periodicals could sometimes contain papers of the highest importance is demonstrated by the occurrence of Mendel's famous paper of 1865 in the *Verhandlungen* of the Naturforscher Verein in Brünn (0641). The long neglect of Mendel's work was not due to any obscurity of the Brünn *Verhandlungen* which, as historical investigations have shown, was available fairly widely (Dorsey, 1944; Weinstein, 1977).

There were few outstanding non-institutional periodicals. The historical significance of *Der Naturforscher* has already been discussed (p. 120). The biggest was the *Annals and Magazine of Natural History* (0614) which was formed in 1841 by a merger of the *Magazine of Natural History* (0605) and the *Annals of Natural History* (0614); the latter had been established in 1838 by Richard Taylor who was the under-secretary of the Linnean Society as well as being the editor and publisher of the *Philosophical Magazine* (p. 124). The *Annals and Magazine of Natural History*, which was almost entirely zoological by the end of the century, has continued to the present and is chiefly concerned with systematics. The second major non-institutional journal was the *Annales des Sciences Naturelles* (0602); it began in 1824 and ten years later split into two sections, one for botany (0689) and one for zoology (0758). Apart from three bibliographical items (0657, 0660, 0661) the other non-institutional journals were quite minor.

MICROSCOPY

Nineteenth-century periodicals for microscopy were of concern to two different groups: on the one hand the amateurs whose interest in the subject was akin to the amateur interest in natural history, and on the other the professionals who were involved in the improvement of the microscope and its applications in anatomy, metallurgy, petrology, and other fields. The two German non-institutional journals begun in the last years of the century (0679 and 0680) were clearly professional in character and it is significant that one of them declared in its title that it was for *wissenschaftliche* microscopy. The provincial society (0678) on the other hand was presumably largely or entirely amateur. The *Quarterly Journal of Microscopical Science* (0675) appears to have been largely amateur in its early years and to have become more professional later in the century. The three societies (0674, 0676, 0677) presumably contained the two elements in varying proportions.

BOTANY

Unlike chemistry and geology, botany was from the earliest times a well-defined science, and by the late eighteenth century it had a considerable literature. The social support which in earlier centuries it had derived from pharmacy, and which had led to the creation of the first botanic gardens and a number of university chairs in the subject, had been, with the obsolescence of herbal pharmacy, largely transferred to chemistry, but in any case the study of plants had sufficient appeal to ensure that the science was well cultivated -- and especially so when the opening up of new regions around the world provided an abundance of interesting new species, some of which also proved to be of practical value in horticulture or agriculture.

There were few botanical periodicals of importance in the early nineteenth century, doubtless because most botanical papers were published in the natural history periodicals or the general science periodicals. Also the book literature appears to have been relatively more important in botany than in most other sciences (Brown, 1956, p. 121). Only two botanical societies were formed before 1850 -- the Bayerische Botanische Gesellschaft in Regensburg (0685) in 1790, and the Botanical Society of Edinburgh (0692) in 1836. The Regensburg Society was very active in the early part of the century (Scudder gives a list of its publications) and its chief periodical, *Flora*, continued into the twentieth century as one of the main German periodicals for the subject. Other significant German items of the period were *Linnaea* (0687) and the *Botanische Zeitung* (0698). There was only one French journal but it was a major one -- the botany section of the *Annales des Sciences Naturelles* (0689). In England, apart from the special case of Curtis' *Botanical Magazine* (0682), there were only minor periodicals and the most important vehicle for the subject was the Linnean Society's *Transactions* (0595) which covered zoology as well. W.J. Hooker's attempts to establish a viable journal (0688) show how difficult it could be, even for a botanist so well placed.

Despite the paucity of important periodicals in the early part of the century there must have been a substantial body of literature in the subject for there were three bibliographical periodicals in existence before 1850. They would of course have covered the botanical papers in the periodicals for natural history and for science in general, and also the book literature. The first of them (0686) was the botanical series in the Stockholm Academy's annual reports on the progress of science, mentioned earlier (p. 150). Like the series for chemistry it was translated into German and was presumably widely read but the Stockholm Academy's scheme as a whole lapsed about mid-century. The other two bibliographical periodicals were separate publications of two of the many annual reports included in the *Archiv für Naturgeschichte* (0759). The first of them (0690), which lasted only from 1837 to 1849, was on the then small field of plant physiology and was initiated by F.J.F. Meyen, a pioneer in the field. The second (0694), which lasted from 1843 to 1857, was on the much wider fields of plant geography and systematic botany and was edited by A. Grisebach, an authority in both fields.

After mid-century some provincial societies were established in Germany (0709, 0711, 0737, 0739) but the Deutsche Botanische Gesell-

Table 47

Botany Periodicals in Existence at Successive Times

	German	British	French	Italian	American	Russian	Other	Total
1790		1					1	2
1800		2					–	2
1810		2					–	2
1820	1	2					–	3
1830	2	2					1	5
1840	3	5	1				2	11
1850	5	7	1	1			3	17
1860	9	7	2	1			3	22
1870	8	7	3	1			6	25
1880	9	9	4	2	2		7	33
1890	12	11	6	3	3		8	43
1900	19	10	7	4	8		11	59

schaft (0728) was not founded until 1882; its *Berichte* soon became of major importance. In Britain the Linnean Society in 1855 began a separate periodical for botany (0707) (as well as one for zoology at the same time) but no significant societies were established, presumably because there were so many natural history societies. The Société Botanique de France (0706) began in 1854 and there was only one French provincial society of any significance (0717). Other countries to set up national societies were Holland (0701), Belgium (0712), Denmark (0715), Japan (0734), and Italy (0741). In the United States there were local societies in New York (0716) and Boston (0751) but a national society was not established until the twentieth century.

Periodicals issued by botanical gardens and herbaria constitute a special category of institutional publications. Most were excluded from the Catalogue because of their smallness; those that were included were from Buitenzorg (in Java) (0724), Kew (0731), Chambésy (in Switzerland) (0744), Berlin (0746), and Geneva (0748).

Specialization into sub-fields is a feature of the botanical literature in the second half of the century, though very limited compared with zoology. Most of the sub-fields corresponded to particular groups of plants -- cryptogams (0703, 0719, 0720), bryophytes (0722, 0750), fungi (0729), ferns (0743), and cacti (0738); there was also a periodical issued by the Académie Internationale de Géographie Botanique (0736).

The largest botanical periodical of the nineteenth century was the British *Journal of Botany* (0713). Presumably much of the British work was concentrated in it because of the paucity of British societies for the subject. Another important, though smaller, British journal was the *Annals of Botany* (0732). The main American periodicals for general botany (0716, 0723, 0749, 0751) were also quite large. Among the German non-institutional journals some of the most outstanding were the *Botanisches Zentralblatt* (0725), Engler's *Botanische Jahrbücher* (0726)--the leading journal in its field of systematics and plant geography--and the *Allgemeine Botanische Zeitschrift* (0747). Especially noteworthy for setting new directions in botany were the *Botanische Zeitung* (0698) and the *Jahrbücher für Wissenschaftliche Botanik* (0708). Other important non-institutional periodicals were the *Revue Générale de Botanique* (0735) and the *Österreichisches Botanisches Wochenblatt* (0702).

Bibliographical periodicals of the late nineteenth century were the *Botanischer Jahresbericht* (0721) and the *Botanisches Zentralblatt* (0725) which became increasingly important as an abstracting journal.

ZOOLOGY

Zoology appears to have been the largest of all the natural sciences during the nineteenth century, the only other science approaching it in size being chemistry. Like botany it had been a well-defined science since ancient times and by the late eighteenth century it had a large body of literature. Even more than in the case of botany the appeal of its subject matter ensured wide social support which found expression in the many natural history societies as well as the purely zoological societies.

Zoology is unique in the degree of its sub-division and the size of its sub-fields, and here too the periodical literature serves as a measure of relative popularity: entomology was the largest of the sub-fields, followed in turn by ornithology, malacology, and ichthyology. As we have seen in the case of natural history, the extent of the amateur interest determines the number of society periodicals, and there were far more of these for entomology, and even more for ornithology, than there were for the much bigger area of general zoology. There was a distinct tendency for the general zoological periodicals to be more professional than those for the sub-fields, but this is not to say that there were no amateurs in general zoology and no professionals in the sub-fields.

(i) General Zoology

In the early decades of the nineteenth century France was the leader in zoology, as it then was in most of the sciences, and its Muséum National d'Histoire Naturelle--which was also an institution for research and teaching--long remained the world's premier institution for the subject. In the first years of the century the Museum's periodical (0753) was the dominant one; it later declined but a new start was made with its *Bulletin* (0800). The other main periodical of the time was also French -- the *Annales des Sciences Naturelles* which in its first period, from 1824, had covered natural history generally (0602) but in 1834 divided into two periodicals, one for zoology (0758) and one for botany. The journals edited by Guérin-Méneville

Table 48

Zoology Periodicals in Existence at Successive Times

(Including sub-fields)

	German	British	French	Italian	American	Russian	Other	Total
1810			1					1
1820	1		1					2
1830	1		1					2
1840	5	4	5					14
1850	7	5	4		1			17
1860	10	8	5		1		3	27
1870	10	10	6	1	3	1	6	37
1880	16	11	8	1	5	1	10	52
1890	22	12	10	2	13	1	17	77
1900	29	18	13	5	19	2	23	109

(0755 and 0760 uniting to become 0764) were also substantial in their early years.

In Britain a national society for the subject was established much earlier than in other countries (as was the case in several other sciences; see p. 138). The Zoological Society of London began in 1826 with the aims of "the advancement of Zoology and Animal Physiology and the introduction of new and curious subjects of the Animal Kingdom." The latter aim was well served by the establishment two years later of the London zoological garden which, besides its well-known functions, constituted a research institution for the Society. In this respect as in others the Society provided a model for later zoological societies. In 1833 it initiated its *Proceedings* (0756) which has continued to the present day as one of the leading periodicals in the field. Another major British periodical originating in the early nineteenth century was *The Zoologist* (0761), and in 1855 the Linnean Society began a separate periodical for zoology (0767) (as well as one for botany at the same time) which, though not large, was important. At about the same time a small periodical was started by the Royal Physical Society of Edinburgh (0766) which dated from 1771 (the word 'physical' in its name had the old meaning) and was now largely or entirely a zoological society.

In Germany in the early years of the nineteenth century the biological sciences (to use a modern expression) were much influenced, for better and for worse, by *Naturphilosophie* which was a particular aspect of the Idealist and Romantic philosophies which were so strong at the time. An outstanding figure in this connection was Lorenz Oken who used his journal *Isis* (0754), much of which he wrote himself, as a means of disseminating the biological ideas of the time. *Isis* however accepted articles from scientists of different opinions and outlooks, encouraged the discussion of problems, and did much to promote research at a time when this was needed in Germany. In the thirty years of its existence the journal served as a national focus for the biological sciences generally. As we have seen earlier (p. 127) Oken also provided an organizational focus for German science.

The *Archiv für Naturgeschichte* (0759), begun in 1835, was a journal of a more orthodox type but with a special and valuable feature in that the latter half of each volume contained annual reviews of many aspects of zoology. The individual reviews continued for varying lengths of time (Scudder gives a list of seventeen which ran for more than one year) and some of them were published separately and continued into the twentieth century (0768 and 0808). Through the nineteenth century and beyond, the *Archiv für Naturgeschichte*, as well as containing papers of the usual type, was a bibliographical journal of the first importance. The fact that such a bibliographical journal could flourish in the early nineteenth century when there were few purely zoological periodicals is a reminder that, as in the case of botany, there was nevertheless a large body of literature in the subject because of the numerous zoological papers that appeared in the periodicals for natural history and for science in general, as well as the book literature which was substantial and, as with botany, more important than in most other sciences.

By the middle of the nineteenth century a rift was opening up between the naturalists, or natural historians, and those who used the relatively new term 'biology' to describe their activities. Hitherto the study of plants and animals had been quite dominated by the descriptive and classificatory approach of the naturalists, but from the early part of the century there was a gradual development of a new concern for the intensive investigation of the vital processes of living organisms, such as respiration, the action of the senses, reproduction, etc. Physiology had hitherto been simply a part of medicine but now there was a move to extend physiological investigations into zoology -- and not for the sake of medicine but for animal physiology as a subject in its own right. The commencement of the *Zeitschrift für Wissenschaftliche Zoologie* (0763) in 1848 by Kölliker and von Siebold, both outstanding biologists, was an assertion of the significance of the new field, and if the term *wissenschaftliche* in its title was interpreted as being rather offensive to the naturalists the biologists would not have minded. The *Zeitschrift* has continued to the present as one of the leading journals for animal biology. (A *Zeitschrift für Wissenschaftliche Botanik* was begun in 1844 by Schleiden and Nägeli but it lasted for only four years and so was not included in the Catalogue. There was however the *Jahrbücher für Wissenschaftliche Botanik* (0808) which began in 1858.)

Apart from the sub-fields of zoology there were not many societies established in the nineteenth century. Besides the pioneering British society, mentioned earlier, there were the Société Zoologique de France (0774) and the Deutsche Zoologische Gesellschaft (0794), founded in 1876 and 1890 respectively, but their periodicals were relatively minor; the German society however took over the important *Zoologischer Anzeiger* (0776) which had been founded twelve years earlier. There were also national societies in Holland (0773), Italy (0792), and Japan (0803). A society was set up in Frankfurt in 1856 but its main concern was the famous zoological garden in that city, and its periodical (0769) dealt with the activities of zoological gardens. Many were established in various cities around the world in the course of the nineteenth century.

Another kind of zoological institution was the natural history museum. The great pioneer was the Paris museum, mentioned earlier, which served as the inspiration and model for many that were set up during the century. Most of them undertook research and published periodicals but many of the periodicals were too small to be included in the Catalogue. Those that were included are the following. (The dates of commencement of their periodicals should not be taken to indicate the museums' dates of foundation.) Albany, N.Y. (0762 and 0786), Klagenfurt (0765 and 0793), Harvard (0770), Tromsø (0775), Leiden (0778), New York (0779), Washington (0780), Vienna (0783), Sydney (0791), Tring, U.K. (0797), St. Petersburg (0801), Capetown (0804). The zoology departments of some universities also issued periodicals in the 1880s and 1890s (0784, 0795, 0796, 0799) but here again many were too small to be included. Yet another kind of institution of relevance to zoology was the U.S. Department of Agriculture (0790).

Leading non-institutional periodicals begun in the late nineteenth century included the *Zoologischer Anzeiger* (0776), founded by the important zoologist J.V. Carus, the *Zoologische Jahrbücher* (0785), and the *Archives de Zoologie Expérimentale et Générale* (0772). As is explicit in the last case their emphasis was chiefly on experimental biology. *The Zoologist* (0761), begun in the early part of the century, continued to be important and with it we can couple the *Annals and Magazine of Natural History* (0614) which by the latter part of the century had become almost entirely zoological.

Being a big science with a rather scattered literature, zoology has a special need for bibliographical services. As we have seen, an important start was made in the early nineteenth century by the *Archiv für Naturgeschichte*, and in 1864 another major bibliographical periodical *The Zoological Record* (0771) was begun. It was instituted by a group of zoologists mostly connected with the British Museum, and early difficulties in funding were resolved when the Zoological Society undertook responsibility for it. From its beginning to the present day *The Zoological Record* has been an indispensable guide to the literature of zoology; though primarily concerned with systematics it can be used for all parts of the subject. Another initiative was the beginning in 1896 of a number of bibliographical serials in card form for zoology and related fields (0533, 0802, 0915, 0918, 0919) by the Concilium Bibliographicum in Zurich. These ceased in 1940, presumably because of wartime difficulties.

(ii) Sub-Fields of Zoology

Entomology was by far the biggest of the sub-fields and its literature is remarkable for the number of society periodicals. From the 1830s onward entomological societies were established in many countries -- France (0806), Britain (0807), Germany (0809, 0812, 0826), Holland (0811), Belgium (0813), Russia (0814), Switzerland (0815), United States (0818, 0821, 0825, 0833), Canada (0819), Italy (0820), and Sweden (0822); in addition there was an international society (0828). Some of the periodicals of the societies in the leading countries were moderately large but the rest were small. There were six non-institutional periodicals (0805, 0816, 0817, 0827, 0830, 0832), none of them outstanding, and a bibliographical periodical (0808) which was a separate publication of one of the series of annual reviews in the *Archiv für Naturgeschichte*.

Ornithology was similar but on a smaller scale, with societies in Germany (0836, 0840, 0842), United States (0841 and 0849), Switzerland (0843), and Britain (0848 and 0851), together with an international society (0845). Some of their periodicals were fairly large. There were five non-institutional journals (0838, 0839, 0846, 0847, 0852), the first of which was the largest.

Malacology (including conchology) had societies in Belgium (0855), Germany (0856), and Britain (0857 and 0860), and four non-institutional journals (0853, 0854, 0858, 0859). For ichthyology there were three periodicals (0861, 0862, 0863) partly representing practical interests in fisheries.

Marine zoology, if it can be called a sub-field, is one very different in character from those just mentioned, being more professional and dependent on special research facilities. The Zoological Station at Naples (0864) was the pioneering research institute and its lead was followed by similar institutions in Britain (0865), Germany (0866), France (0867), and the United States (0868).

EXPERIMENTAL BIOLOGY

The term 'experimental biology' is a twentieth-century term for what in the nineteenth century was simply called biology. Today we generally include botany and zoology under the heading of biology but this was rarely done in the nineteenth century: biology then was something relatively new and quite different in its aims and methods from the traditional botany and zoology. (If today we add the adjective 'experimental' to specify what was then new we have to stretch the meaning of the word somewhat to include fields like histology which generally use observation rather than experiment -- but systematic observation carried on in strictly controlled conditions with the use of instruments, especially the microscope.)

The leading component of experimental biology in the nineteenth century was physiology, and we have already noticed, in connection with the significantly named *Zeitschrift für Wissenschaftliche Zoologie*, the introduction of the physiological approach into zoology. Physiology had hitherto been simply a part, and only a small part, of theoretical medicine -- a minor handmaiden of the healing art. In the course of the nineteenth century it became a major science in its own right, and one which dealt with functions and processes occurring in animals and humans alike. Inseparably connected with physiology was anatomy, the oldest of the medical sciences. By the early years of the nineteenth century gross (or naked-eye) anatomy had been largely worked out but the big improvements in the microscope from the 1830s onward created the new field of microscopic anatomy while at the same time revitalizing the old science of embryology. These developments were as applicable to plant and animal tissues as to human tissues, and later in the century they led to the emergence of the closely related field of cytology which was full of potentialities for the twentieth century.

The first significant periodical to carry the word 'physiology' in its title was the *Archiv für die Physiologie* (0869) which ran from 1795 to 1832 under the editorship first of J.C. Reil and then of J.F. Meckel. Its scope was much wider than its title indicates for as well as much human and comparative anatomy it included many articles of a rather philosophical kind on what we would now call general biology, and even some on chemistry and physics. As we have already seen in the case of Oken and his journal *Isis*, biology in Germany in the first decades of the nineteenth century was greatly influenced by the speculations of *Naturphilosophie*, and though Reil and Meckel tried to uphold the value of observation and experiment the philosophical tide was running against them, especially as the fund of factual knowledge of vital phenomena was then so small. Compared with what was to come in a few decades it was a low period for biology in Germany.

Once again it was France that was the leader in the opening decades of the century, and in France the philosophical climate was utterly different. In scientific and medical circles an extreme form of positivism prevailed and, however destructive such a philosophy may be in other spheres of life, in medicine and the biological sciences at that time it had a purgative effect which proved beneficial. It was the chief influence in the work of François Magendie, the main pioneer of experimental physiology. Magendie's *Journal de Physiologie Expérimentale* (0870), begun in 1821, contained only some of his many discoveries and for some reason ceased in 1831 when he acceded to the chair of medicine at the Collège de France. Much of the physiological research done by him and his successors was published in medical periodicals.

Two biological societies were founded in the first half of the century, both in France. The Société Anatomique (0871) was oriented mainly towards pathological anatomy and medicine, but the Société de Biologie (0876), founded in 1849, firmly established both the word (then still relatively new) and the field of biology in French scientific life. With the rapid development of the field in the late nineteenth century the Société followed the lead of the Académie des Sciences in publishing weekly *Comptes Rendus* containing short communications. As a result its *Comptes Rendus* became the vehicle for most French work in the subject.

From about 1830 the establishment of experimental biology in Germany and--what was to prove decisive--its integration into the developing German university system was largely the work of Johannes Müller. What Liebig was to chemistry Müller was to experimental biology. As well as being an important discoverer he was also a great teacher, and nearly all the leading German biologists of the next generation had been his students. In his time (though not in theirs) it was possible to take the whole field of experimental biology as his province and in 1833 he became professor of anatomy and physiology at the University of Berlin, the apex of the German university system. In the following year he began his *Archiv für Anatomie, Physiologie und Wissenschaftliche Medicin* (0872) which immediately became the

Table 49

Experimental Biology Periodicals in Existence at Successive Times

	German	British	French	Italian	American	Russian	Other	Total
1800	1							1
1810	1							1
1820	1							1
1830	1		2					3
1840	1		1					2
1850	3		2				1	6
1860	6		3	1			1	11
1870	12	1	4	1			1	19
1880	11	3	4	2			1	21
1890	16	4	5	3	2		4	34
1900	20	4	9	3	4	2	7	49

main vehicle for German work in the field. When he died in 1858 his chair was divided: C.B. Reichert received the professorship of anatomy and E. Du Bois-Reymond that of physiology, and the *Archiv* continued under their joint editorship.

Another German periodical of the time which was important for experimental biology though it was primarily medical was Virchow's *Archiv für pathologische Anatomie und Physiologie* (0874). Virchow was a pathologist who made a major contribution to pure biology by his successful advocacy of the cell theory. Also appearing in a medical context were two bibliographical periodicals (0873 and 0877); the former ceased after thirty years but the latter continued into the twentieth century as a major review serial for the whole field. Virchow also edited a *Jahresbericht* for anatomy and physiology (0888). In the 1850s and 1860s the growing independence of physiology from medicine is illustrated by the practice of a number of periodicals (0878, 0881, 0882, 0889) making the point in their titles that they covered animal as well as human physiology.

The rapidly developing importance of microscopic anatomy was demonstrated by the appearance in 1865 of the *Archiv für Mikroskopische Anatomie* (0883), edited by the eminent anatomist and cytologist M. Schultze, which became the leading journal in the field. At about the same time an important *Archiv* for general physiology began (0889), edited by the versatile physiologist, F.W. Pflüger. "He exercised a strong influence on his contemporaries while serving as editor of *Archiv für gesammte Physiologie des Menschen und der Thiere* from 1868. Few people went as far as Pflüger in public criticism of the works of others, and he was the most feared critic of his time." (DSB) After his death in 1910 his journal was renamed *Pflüger's Archiv....* and it has continued to the present as one of the leading journals.

Another physiological journal appearing in the 1860s was entitled significantly *Zeitschrift für Biologie* (0884). A few years later the advent of a new specialty within physiology, namely pharmacology, was attested by the establishment of a journal for the subject (0891). Another journal begun in the 1870s (0892) was for the very old subjects (both of which went back to Aristotle) of comparative anatomy and embryology, now transformed by the use of the microscope; it was edited by the eminent anatomist, C. Gegenbauer.

A feature of the highly fruitful integration of experimental biology into the German university system was the establishment by the leading universities of special institutes for research and advanced teaching in the field. One of the most famous of these was the institute constructed by the University of Leipzig for the great physiologist Carl Ludwig when he accepted the newly created chair of physiology in 1865. Ludwig's institute began its own periodical (0885) as did the laboratory of Adolf Fick, professor of physiology at Würzburg (0886), and there were examples of similar periodicals in other countries (0875, 0890, and other items excluded because of their smallness). Periodicals of this type however did not become of any importance and the research results of the institutes continued to be published in the usual periodicals.

Experimental biology in France remained substantial though from about the 1840s it became overshadowed by the great expansion taking place in Germany. Three significant new periodicals were established (0881, 0882, 0921) as well as two bibliographical items (0914 and 0917) but, as mentioned earlier, in the late nineteenth century most of the French work in the field was concentrated in the *Comptes Rendus* of the Société de Biologie; as a result by the end of the century, if not earlier, it was the biggest of all periodicals in the field in terms of the number of papers.

Experimental biology made remarkably little progress in Britain until the 1870s, possibly because medicine was mostly taught outside the universities and because Oxford and Cambridge gave little support to either medicine or science (with the exception of mathematics at Cambridge) until then. The first sign of improvement was the beginning in 1867 of the *Journal of Anatomy* (0887) followed in 1870 by the appointment of Michael Foster to a post in physiology at Cambridge. Foster was a great teacher with an enthusiasm for original research which inspired a brillant band of students, and by the 1880s he had built up an important school of physiology at Cambridge. He was also the main influence in the establishment of the Physiological Society as well as being the editor of the *Journal of Physiology* (0893) which began in 1878. The neurological journal *Brain* (0894) also began in the same year.

In the United States likewise there were few developments until the establishment in 1876 of the Johns Hopkins University which was conceived as a research-oriented graduate school with the combination of teaching and research which had long been a leading characteristic of the German universities but was still unfamiliar in other countries. The new university had an emphasis on biology and one of Foster's disciples at Cambridge, H.N. Martin, was appointed professor and he too built up a research school and was instrumental in creating the American Physiological Society in 1887. The *American Journal of*

Physiology (0922), published by the Society, began in 1898. The Biological Society of Washington began its *Proceedings* (0895) in 1883, followed a few years later by the *Journal of Morphology* (0906) and the *Journal of Comparative Neurology* (0910). In the twentieth century the English-speaking countries made up for their late start, and today both the *Journal of Physiology* and the *American Journal of Physiology* are among the leading periodicals in the field.

In Italy the *Archivio Italiano di Biologia* (0898--generally in French, perhaps because it emanated from Turin) was begun in 1882 by A. Mosso, professor of physiology at Turin and head of a flourishing research institute; by the end of the century it had become a major periodical. There was also a small Italian periodical (0879) which had begun much earlier. The *Skandinavisches Archiv für Physiologie* (0907) was established in 1889 by Sweden's first professor of physiology, F. Holmgren. In Russia two biological periodicals (0913 and 0920) began in the 1890s, both issued by governmental laboratories. There was also an international journal for anatomy and physiology (0900), published from 1884 in both German and English.

Societies did not play the same part in experimental biology that they did in some other sciences (with the notable exception of the Société de Biologie), and when the Physiologische Gesellschaft zu Berlin was established in 1875 it did not institute a periodical of its own, doubtless because there were already so many in the field, but instead gave its support to a bibliographical journal, the *Centralblatt für Physiologie* (0905). The Anatomische Gesellschaft, however, at its foundation in 1886 began the *Anatomischer Anzeiger* (0903) which covered the whole field of anatomy and soon became one of the biggest journals in experimental biology.

Meanwhile non-institutional journals continued to appear in Germany. In the last two decades of the century the biggest to be established were the *Biologisches Centralblatt* (0897), the *Anatomische Hefte* (0911), and the *Archiv für Entwicklungsmechanik* (0916). The last was edited by the notable embryologist Wilhelm Roux and was for many years the leading journal in its field. The *Morphologische Arbeiten* (0912) was smaller but significant. In addition there were a number of periodicals in the border region between biology and medicine. Those that were included in the Catalogue are *Neurologisches Zentralblatt* (0899), *Beiträge zur Pathologischen Anatomie und Physiologie* (0901), and *Centralblatt für Allgemeine Pathologie und Pathologische Anatomie* (0909). There were also the periodicals for biochemistry (0423 and 0427) and food chemistry (0436 and 0447) which have been placed under Chemistry.

The growth of experimental biology during the nineteenth century was prodigious. At the beginning of the century physiology was confined to medicine as a small appendage of anatomy, while at the end it had become one of the biggest of the natural sciences. And the growth in the various anatomical and morphological fields was comparable. An indication of size is provided by the bibliographical periodicals in existence at the end of the century. Besides the two previously mentioned which had begun earlier (0877 and 0888), the last two decades saw the establishment of two more German items (0897 and 0905), three French items (0914, 0915, 0917), and two items in the international series conducted by the Concilium Bibliographicum (0918 and 0919).

MICROBIOLOGY

The history of microbiology falls into two very different periods divided from each other by the epoch-making discoveries of Pasteur and Koch in the late 1870s which established the germ theory of disease. The earlier period extends back to microscopists like Leeuwenhoek in the seventeenth century and culminates in the researches beginning about 1850, of the German botanist F.J. Cohn. From 1870 Cohn published his results in the small journal he had founded under the name of *Beiträge zur Biologie der Pflanzen* (0924). (Bacteria were then regarded as primitive plants and their study as part of botany.) Before 1880 Cohn's work was academic in every sense of the word but with the general realization thereafter of the vast significance of the germ theory of disease the subject of bacteriology was utterly transformed.

Before the discoveries of Pasteur and Koch a different approach had been under way in an area which today we would call epidemiology but which was then included in what was called hygiene -- a subject concerned with air, water, food, sanitation, etc., and located somewhere between chemistry and medicine. From the 1860s chairs of hygiene existed at several German universities and the subject was largely the creation of Max Pettenkofer who established the first periodical for it in 1883 (0925). In the same year a *Jahresbericht* (0926) began; most of the literature in the subject was doubtless in the periodicals for medicine and chemistry. The germ theory of disease provided not only a foundation for "hygiene" but a different way of looking at its problems, and though the term was still used in 1886 for a new periodical (0928) by none other than Koch, it gradually disappeared as the name of a subject.

After 1880 the field of microbiology--then almost entirely bacteriology--expanded at an extraordinary rate and the periodicals listed in the Catalogue represent only the beginning of a virtual explosion in the literature of the field, indeed only a part of the beginning because they do not include the medical periodicals, except for some borderline cases.

APPENDIX

The Periodical Literature of the Individual Sciences at the Beginning of the Twentieth Century

As described in Section 2.2, one of the surveys undertaken to estimate the sizes of the periodicals made use of the *International Catalogue of Scientific Literature* for the period 1901-05 because this catalogue is sub-divided by subjects. The figures initially obtained from it were counts of references within each subject category; these were then collated to give the total counts (for all subjects) for each periodical, and thus the size of the periodical. Here we are concerned, not with the sizes of the periodicals, but with the initial figures as measures of the periodical literatures of the individual sciences. The figures were in the form of average numbers of references per year during the period 1901-05 and were obtained by the sampling procedure described in Section 2.2.

For each of the subject categories of the *International Catalogue* the periodicals cited in it were ranked in decreasing order of the number of references. The list for each subject was long, the number of periodicals ranging from 152 for mathematics to 594 for zoology, and the lists need not be reproduced here. In their general import they were similar to Tables 21 to 33 but the list for each subject included periodicals for science in general and periodicals for related sciences. Within each list the distribution of the references followed a curve of the inverse-square type, with a small proportion of high-ranking periodicals accounting for most of the references and a long "tail" of minor periodicals with very few references each.

Of chief interest are the total numbers of references for the various subjects. These are set out in Table 50 using the subject categories of the *International Catalogue* which are somewhat different from those used in Part 1 of the present work. (It also has a section for Physical Anthropology which was not counted.) The *International Catalogue* is very widely inclusive and in most of its sections it contains many references of an applied or practical nature. These were excluded from the count but it was often difficult to decide where the borderline lay. This was especially the case with mechanics, physics, chemistry, and botany, and most of all with the three sciences bearing on medicine -- human anatomy, physiology, and bacteriology. Consequently several of the totals are rather indefinite but the figures in Table 50 should be good enough to give an approximate indication of the relative sizes of the periodical literatures of the individual sciences.

Table 50

Average Number of References Per Year in 1901-05

		%
Mathematics	1185	3.9
Mechanics	528	1.7
Physics	2556	8.5
Chemistry	4986	16.5
Astronomy	1439	4.8
Meteorology	983	3.3
Mineralogy	1038	3.4
Geology	1218	4.0
Geography	1146	3.8
Palaeontology	541	1.8
General Biology	582	1.9
Botany	3423	11.3
Zoology	5316	17.6
Human Anatomy	1166	3.9
Physiology	3285	10.9
Bacteriology	828	2.7
	30,220	100.0

The introduction to the last volume of the Royal Society's *Catalogue of Scientific Papers* (Vol. XIX, published 1925) gives a total of 384,478 for the number of papers cited in the Fourth Series of the *Catalogue*, covering the years 1884-1900 inclusive. This is equivalent to an average of 22,616 per year during the period, a figure which is in line with the above total if allowance is made for the continuous exponential increase.

BIBLIOGRAPHY

Académie des Sciences. 1910-13. *Procès-verbaux des séances de l'Académie tenues depuis la fondation de l'Institut jusqu'au mois d'août 1835*. 4 vols. Paris.

Andrade, E.N. 1965. The birth and early days of the *Philosophical transactions*. *Notes and Records of the Royal Society* 20: 9-27.

Barnes, S.B. 1934. The scientific journal, 1665-1730. *Scientific Monthly* 38: 257-260.

Barnes, S.B. 1936. The editing of early learned journals. *Osiris* 1: 155-172.

Barr, E.S. 1964. Biographical material in the *Philosophical magazine* to 1900. *Isis* 55: 88-90.

Bauer, M., et al. 1907. Neues Jahrbuch für Mineralogie, Geologie und Paläontologie. *Festband zur Feier des 100 jährigen Bestehens*. Stuttgart.

Beilstein, F.K. 1972. *Beilstein-Erlenmeyer: Briefe zur Geschichte des chemischen Dokumentation und des chemischen Zeitschriftwesens*. Munich.

Besterman, T. 1945. A bibliography of the *Bulletin* of the Academy of Sciences of the U.S.S.R. *J. Documentation* 1: 45-56.

Bickerton, D.M. 1972. A scientific and literary periodical, the *Bibliothèque britannique* (1796-1815). *Revue de Littérature Comparée* 46: 527-547.

Birn, R. 1965. Le *Journal des savants* sous l'ancien régime. *J. des Savants* 1965: 15-35.

Blake, D., and R.E.M. Bowden. 1968. Journal of anatomy: *Index to the first 100 years, 1866-1966*. Cambridge.

Blunt, W. 1950. *The art of botanical illustration*. London. Chapter 15: "The *Botanical magazine*."

Boig, F.S., and P.W. Howerton. 1952a. History and development of chemical periodicals in the field of organic chemistry, 1877-1949. *Science* 115: 25-31.

Boig, F.S., and P.W. Howerton. 1952b. History and development of chemical periodicals in the field of analytical chemistry, 1877-1949. *Science* 115: 555-559.

Bolton, H.C. 1897. *Catalogue of scientific and technical periodicals, 1665-1895*. 2nd ed. Washington. Reprinted, New York, 1965. Does not include institutional periodicals.

Brown, Charles Harvey. 1956. *Scientific serials: Characteristics and lists of most cited publications*. Chicago. "A classic study."

Brown, Harcourt. 1934. *Scientific organizations in seventeenth-century France*. New York. Chapter IX: "Science and the press."

Brown, Harcourt. 1972. History and the learned journal [*Journal des sçavans*]. *J. Hist. Ideas* 33: 365-378.

Court, S. 1972. The *Annales de chimie*, 1789-1815. *Ambix* 19: 113-128.

Crosland, M. 1967. *The Society of Arcueil: A view of French science at the time of Napoleon I*. London.

Dana, E.S., et al. 1918. *A century of science in America, with special reference to the* American journal of science *from 1818-1918*. New Haven.

Delépine, M. 1962. Les *Annales de chimie*. *Annales de Chimie*, 13e sér., 7: 1-11.

Desautels, A.R. 1956. *Les* Mémoires de Trévoux *et le mouvement des idées au XVIII*^{e} *siècle, 1701-1734*. Rome.

Dorsey, M.J. 1944. Appearance of Mendel's paper in American libraries. *Science* 99: 199-200.

(DSB) *Dictionary of scientific biography*. 14 vols with supplement and index. New York, 1970-80.

Dulieu, L. 1958. La contribution montpelliéraine aux recueils de l'Académie Royale des Sciences. *Rev. Hist. Sci*. 11: 250-262.

Dumas, G. 1936. *Histoire du* Journal de Trévoux *depuis 1701 jusqu'à 1762*. Paris.

Elkhadem, H. 1978. Histoire de la *Correspondance mathématique et physique* d'après les lettres de J.G. Garnier et A. Quetelet. *Bull. Acad. Roy. Belgique* 64: 316-366.

Elliott, C.A. 1970. The Royal Society "Catalogue" as an index to nineteenth-century American science. *J.Amer. Soc. Inf. Sci*. 21: 396-401.

Ferguson, A. (ed.) 1948. *Natural philosophy through the eighteenth century*. London. Reprinted 1972. Re the sesqui-centenary of the *Philosophical magazine*.

Fifield, D. 1969. [The centenary of *Nature*] *New Scientist*, 30 Oct. 1969, pp. 230-232.

Freund, H. 1964. 80 Jahre *Zeitschrift für wissenschaftliche Mikroskopie*. *Zeit. wiss. Mikroskopie* 66: 3-17.

Garrison, F.H. 1934. The medical and scientific periodicals of the seventeenth and eighteenth centuries. *Bull. Inst. Hist. Med*. 2: 285-343.

Gascoigne, R.M. 1984. *A historical catalogue of scientists and scientific books*. New York.

Gauja, P. 1934. *L'Académie des Sciences*. Paris.

George, P. 1952. The scientific movement and the development of chemistry in England, as seen in the papers published in the *Philosophical transactions* from 1664/5 until 1750. *Annals Sci*. 8: 302-322.

Geus, A. (ed.) 1971. *Indices naturwissenschaftlich-medizinischer Periodica. Bis 1850.* [A series published at Stuttgart from 1971 onward] Vol. 1: Der Naturforscher *(1774-1804).* Vols 2 and 3: *Die chemischen Zeitschriften des Lorenz von Crell (1778-1804).* Vol. 4: *Die Schriften der Gesellschaft Naturforschender Freunde zu Berlin (1775-1838).* Vol. 5: *Die medizinischen Zeitschriften des J.F. Meckel.* [Includes *Deutsches Archiv für die Physiologie*, 1815-23, and *Archiv für Anatomie und Physiologie*, 1826-32]

Gregory, R. 1943. News and views from the scientific front. [Re *Nature*] *Nature* 151: 517-519.

Hall, M.B. 1965. Oldenburg and the art of scientific communication. *British J. Hist. Sci.* 2: 277-290.

Hall, M.B. 1975. The Royal Society's role in the diffusion of information in the seventeenth century. *Notes and Records of the Royal Society* 29: 173-192.

Harff, H. 1941. *Die Entwicklung der deutschen chemischen Fachzeitschrift.* Berlin.

Herrmann, D.B. 1971. B.A. Gould and his *Astronomical journal*. *J. Hist. Astron.* 2: 98-108.

Herrmann, D.B. 1972. *Die Entstehung der astronomischen Fachzeitschriften in Deutschland (1798-1821).* Berlin-Treptow. (Veröffentlichungen der Archenhold-Sternwarte. Nr. 5)

Herrmann, D.B. 1973. An exponential law for the establishment of observatories in the nineteenth century. *J. Hist. Astron.* 4: 57-58.

Houghton, B. 1975. *Scientific periodicals: Their historical development, characteristics and control.* London.

Hückel, W. 1967. 100 Jahre Geschichte der *Berichte* [der Deutschen Chemischen Gesellschaft] *Chem. Berichte* 100: I-LIII.

Hufbauer, K. 1971. Social support for chemistry in Germany during the eighteenth century. *Hist. Stud. Phys. Sci.* 3: 205-231.

Hunkin, J.W. 1946. William Curtis, founder of the *Botanical magazine*. *Endeavour* 5: 13-17.

International catalogue of scientific literature. London. 1st-14th annual issue, 1902-1917.

Jaeggli, A.E. 1977. Recueil des pièces qui ont remporté le prix de l'Académie Royale des Sciences. *Gesnerus* 34: 408-414.

Knight, D. 1975. *Sources for the history of science, 1660-1914.* London. Chapter 4: "Journals"

Kronick, D.A. 1958. The Fielding H. Garrison list of medical and scientific periodicals of the seventeenth and eighteenth centuries. Addenda et corrigenda. *Bull. Hist. Med.* 32: 456-474.

Kronick, D.A. 1965. Scientific journal publication in the eighteenth century. *Papers Bibliog. Soc. America* 52: 28-44.

Kronick, D.A. 1976. *A history of scientific and technical periodicals, 1665-1790.* 2nd ed. Metuchen, N.J.

Kronick, D.A. 1978. Authorship and authority in the scientific periodicals of the seventeenth and eighteenth centuries. *Library Quarterly* 48: 255-275.

Lilley, S. 1948. Nicholson's Journal (1797-1813). *Annals Sci.* 6: 78-101.

Loria, G. 1899. Il *Giornale de' letterati d'Italia* e la *Raccolta Calogera* come fonti per la storia delle matematiche nel secolo XVIII. *Abhandlungen zur Geschichte der Mathematik* 9: 243-274.

Loria, G. 1941. *Acta eruditorum* durante gli anni 1682-1740 e la storia delle matematiche. *Archeion* 23: 1-35.

Lyons, H. 1944. *The Royal Society, 1660-1940*. Cambridge.

Maas, K. 1976. 100 Jahre *Chemiker Zeitung*. *Chem. Zeitung* 100: 155.

McClellan, J.E. 1979. The scientific press in transition: Rozier's Journal and the scientific societies in the 1770s. *Annals Sci.* 36: 425-449.

McKie, D. 1948. "The scientific periodical from 1665 to 1798." *Natural philosophy through the eighteenth century*. Ed. by A. Ferguson. London, 1948 (reprinted 1972). pp. 122-132.

McKie, D. 1957. The *Observations* of the Abbé François Rozier. *Annals Sci.* 13: 73-89.

MacLeod, R.M. 1969. [Articles on the history of *Nature* on the occasion of its centenary] *Nature* 224: 423-461.

Meadows, A.J. 1974. *Communication in science*. London. Chapter 3: "The rise of the scientific journal".

Morgan, B.T. 1929. *Histoire du* Journal des sçavans *depuis 1665 jusqu'à 1701*. Paris.

Müller, Johannes. 1883-1917. *Die wissenschaftlichen Vereine und Gesellschaften Deutschlands im neunzehnten Jahrhundert: Bibliographie ihrer Veröffentlichungen*. 2 vols. Berlin. (Reprinted Hildesheim, 1965)

Nature. 1969. [Editorial re the centenary of *Nature*] *Nature* 224: 417-422.

Neave, E.W.J. 1950-1952. Chemistry in Rozier's Journal. *Annals Sci.* 6: 416-421; 7: 101-106, 284-299, 393-400; 8: 28-45.

Neave, S.A. 1950. The *Zoological record*. *Science* 112: 761.

Ornstein, M. 1938. *The role of scientific societies in the seventeenth century*. 3rd ed. Chicago.

Osborn, H.F. 1903. Progress of the Concilium Bibliographicum. *Science* 17: 951-952.

Paris, G. 1903. Le *Journal des savants*. *J. des Savants*, Jan. 1903, pp. 5-34.

Pflücke, M. 1954. Das *Chemische Zentralblatt* 125 Jahre alt. *Angew. Chem.* 66: 537-541.

Phillips, J.P. 1966. Liebig and Kolbe, critical editors. *Chymia* 11: 89-97.

Poggendorff, J.C. 1863. *Biographisch-literarisches Handwörterbuch zur Geschichte der exacten Wissenschaften*. Vols I and II. Leipzig, 1863; reprinted Amsterdam, 1965. (The subsequent volumes were not used in the present work.)

Ranc, A. 1949. L'évolution de la presse de diffusion des sciences depuis le début du XIXe siècle. *Sciences* 76: 87-107.

Reuss, J.D. 1801-21. *Repertorium commentationum a societatibus literariis editarum.* 16 vols. Göttingen, 1801-21; reprinted New York, 1961.

Roepke, W. 1928. Die Veröffentlichungen der Kaiserlich Leopoldinisch Deutschen Akademie der Naturforscher. *Leopoldina* 3: 149-158.

Rothschuh, K.E., and A. Schäfer. 1955. Quantitative Untersuchungen über die Entwicklung des physiologischen Fachschrifttums (Periodica) in den letzten 150 Jahren. *Centaurus* 4: 63-66.

Royal Society of London. 1867. *Catalogue of Scientific Papers.* [1800-1900] 19 vols and subject index of 3 vols in 4. London, 1867-1925.

Royal Society of London. 1912. *Catalogue of the periodical publications in the library.* London.

Russo, F. 1969. *Eléments de bibliographie de l'histoire des sciences et des techniques.* 2nd ed. Paris.

Sarton, G. 1944. Vindication of Father Hell. *Isis* 35: 97-105.

Sarton, G. 1952. *Guide to the history of science.* Waltham, Mass.

Scheler, L. 1961. Lavoisier et le *Journal d'histoire naturelle. Rev. Hist. Sci.* 14: 1-9.

Schimank, H. 1963. L.W. Gilbert und die Anfänge der *Annalen der Physik. Sudhoffs Archiv* 47: 360-372.

Scudder, S.H. 1879. *Catalogue of scientific serials, 1633-1876.* Cambridge, Mass., 1879; reprinted New York, 1965.

Sergescu, P. 1936. Les mathématiques dans le *Journal de sçavans*, 1666-1701. *Osiris* 1: 568-583.

Sergescu, P. 1947. La littérature mathématique dans la première période (1665-1705) du *Journal des sçavans. Archives Internat. Hist. Sci.* 1: 60-99.

Sheets-Pyenson, S. 1981a. Darwin's data: His reading of natural history journals, 1837-1842. *J. Hist. Biol.* 14: 231-248.

Sheets-Pyenson, S. 1981b. A measure of success: The publication of natural history journals in early Victorian Britain. *Publishing History* 9: 21-36.

Silliman, R.H. 1974. Fresnel and the emergence of physics as a discipline. *Hist. Stud. Phys. Sci.* 4: 137-162.

Smeaton, W.A. 1957. *L'Avant-coureur*: The journal in which some of Lavoisier's earliest work was reported. *Annals Sci.* 13: 219-234.

Taton, R. 1947. Les mathématiques dans le Bulletin de Férussac. *Archives Internat. Hist. Sci.* 1: 100-125.

Thornton, J.L., and R.I.J. Tully. 1971. *Scientific books, libraries and collectors.* 3rd ed. London. Chapter VIII: "The growth of the scientific periodical literature".

Thornton, J.L., and R.I.J. Tully. 1978. *Scientific books, libraries and collectors. Supplement 1969-75.* London. Chapter VIII.

Van Klooster, H.S. 1957. The story of Liebig's *Annalen der Chemie*. *J. Chem. Education* 34: 27-30.

Walls, E.W. 1966. The *Journal of anatomy*, 1867-1966. *J. Anatomy* 100: 1-4.

Weinstein, A. 1977. How unknown was Mendel's paper? *J. Hist. Biol.* 341-364.

Wightman, W.P.D. 1961. *Philosophical transactions* of the Royal Society. *Nature* 192: 23-24.

Willstätter, R. 1929. Zur Hundertjahrfeier des *Chemischen Zentralblatts*. *Zeit. Angew. Chem.* 42: 1049-1052.

Yagello, V.E. 1968. Early history of the chemical periodical. *J. Chem. Education* 45: 426-429.

INDEX

Including titles of periodicals and names of persons, institutions, and cities. Institutional periodicals are entered in the name of their respective institution if, as is usually the case, the institution's name forms part of the periodical's title (thus institutional periodicals are not entered under such titles as *Berichte*, *Bulletin*, *Mitteilungen*, *Proceedings*, *Transactions*, etc., except in a few special cases). There are duplicate entries for many institutions under the name of the city in which each was located. An institution can often be found most easily by looking under the name of the city.

Numbers beginning with 0 are entry numbers in Part 1. Other numbers are page numbers.

Aardrijskund. Genootsch. 0559
Abbe, E. 0380
Abeille 0817
Abh. Geschichte Math. 0304
Abh. Privatgesellsch. Böhmen 033; 94
Acad. Caes. Leopold. Nat. Curios. *See* Acad. Nat. Curios.
Acad. Ciencias Madrid 0142
Acad. Elector. Mogunt. Sci. etc. 024; 93
Acad. Electór. Sci. etc. Mannheim 028; 93
Acad. Internat. Géogr. Botan. 0736
Acad. Nat. Curios. 03, 0705; 93, 97, 116, 127
Acad. Nat. Sci. Philadelphia 0598, 0616, 0831; 106, 111
Acad. Sci. Berlin 07, 017, 0356; 93, 97, 118, 120
Acad. Sci. etc. Bordeaux 09; 93
Acad. Sci. etc. Bruxelles 034, 073, 093; 94, 103
Acad. Sci. etc. Clermont-Ferrand 089
Acad. Sci. Erford. 024
Acad. Sci. Lisboa 0203
Acad. Sci. etc. Lyon 0121
Acad. Sci. etc. Montpellier 0131
Acad. Sci. Paris *(to 1793)* 05, 010, 018, 0351; 93, 116, 117, 119, 121, 141, 158
——— *(Histoire et) Mémoires* 05; 7, 93, 97, 117, 119, 121,122
Acad. Sci. Paris *(from 1816)* 066, 088, 098; 122, 123
——— *Comptes rendus* 098; 103, 107, 122-123, 131, 137, 140, 143, 151, 166
Acad. Sci. Petropol. 012; 94, 97. *See also* Acad. Sci. St. Pétersbourg
Acad. Sci. St. Louis 0164
Acad. Sci. St. Pétersbourg 055, 0101. *See also* Acad. Sci. Petropol.
Acad. Sci. etc. Toulouse 036, 087; 93, 121
Acad. Sci. Turin. *See* Accad. Sci. Torino
Acad. Stanislas 054
Acad. Theodoro-Palatina 028
Accad. Cimento 116
Accad. Fisiocritici 026
Accad. Gioenia Sci. Nat. Catania 085, 0255
Accad. Lincei 0211; 107, 131
Accad. Medico-Fis. Fiorentina 0879
Accad. Pontif. Nuovi Lincei 0132, 0211, 0252
Accad. Sci. etc. Bologna 090, 0140
Accad. Sci. etc. Napoli 070, 0115, 0174, 0175; 103

Accad. Sci. etc. Padova 042; 94, 121
Accad. Sci. Siena 026; 94
Accad. Sci. Torino 038, 065, 0189; 94, 97
Acta erudit. 04; 93, 98, 116-117
Acta Helvet. 019; 94
Acta Leipsiensia. See *Acta erudit.*
Acta lit. Sueciae 011
Acta math. 0307
Acta med. philos. Hafniensia 95
Acta Petropolitana 012
Actual. chim. 0453
Adansonia 0710; 106
Adelaide
  Philos. Soc. 0156
  Royal Soc. South Australia 0156
Afhandl. fys. kem. mineral. 0849
Agassiz Assoc. 0849
Agence des Mines 0455; 153
Ahrens 0452
Akad. Naturforscher. *See* Acad. Nat. Curios.
Akad. Nauk St. Peterburg 0101, 0801
Akad. Umiej. Krakow 0224, 0259; 107
Akad. Wetensch. Amsterdam 0155, 0266; 107, 131
Akad. Wissensch. Berlin 056, 0103; 103, 107, 131
Akad. Wissensch. Göttingen. *See* Gesellsch. Wissensch.
Akad. Wissensch. Leipzig. *See* Sächs. Gesellsch. Wissensch.
Akad. Wissensch. München. *See* Bayer. Akad. Wissensch.
Akad. Wissensch. Wien 0135, 0141, 0146, 0184, 0431, 0537; 103, 107, 131
Åkerman, J. 0458
Albany, N.Y.
  State Cabinet Nat. Hist. 0762
Alembert, J. d' 117
Alessandrini, A. 0615
Algemeem mag. wetensch. etc. 97
Allg. berg- und huttenmänn. Zeit. 0471
Allg. botan. Zeitschr. Systematik etc. 0747; 112, 161
Allg. Chemiker Zeit. 0428
Allg. Conferenz Internat. Erdmessung 0572
Allg. deutsch. naturhist. Zeit. 0622
Allg. Entomol. Gesellsch. 0835
Allg. Fischereizeit. 0861
Allg. Journ. Chem. 0397; 149
Allg. Schweiz. Gesellsch. gesammt. Naturwissensch. 086
Allg. Zeitschr. Entomol. 0835; 113
Allis, E.P. 0906
Almanaque náutico etc. 0357
Almanach Scheidekünst. Apoth. 0392a; 99
Almeida, J.C.d' 0370; 146
Altona. gelehr. Anzeigen 96
Amer. Acad. Arts Sci. 041, 0127; 94, 121
Amer. Assoc. Adv. Sci. 0136, 0244; 153
Amer. chem. journ. 0429; 151
Amer. Chem. Soc. 0430; 109, 151
Amer. Entomol. Soc. 0818
Amer. ephem. naut. almanach 0361
Amer. Fern Soc. 0743
Amer. Geogr. (Stat.) Soc. 0547; 105
Amer. geologist 0516; 110, 155
Amer. journ. math. 0305
Amer. journ. pharm. 0406; 105
Amer. journ. physiol. 0922; 113, 168-169
Amer. journ. sci. 069; 103, 107, 125, 130-131, 154, 155
Amer. math. monthly 0320; 108
Amer. Math. Soc. 0317
Amer. Microscop. Soc. 0677
Amer. museum 97
Amer. Museum Nat. Hist. 0779
Amer. naturalist 0642; 111
Amer. Philos. Soc. 029, 0106; 94, 121
Amer. Physical Soc. 0388; 146
Amer. Physiol. Soc. 0922; 168
Amer. Soc. Microscop. 0677
Amsterdam
  Aardrijskund. Genootsch. 0559
  Akad. Wetensch. 0155, 0266; 107, 131
  Soc. Math. 0319
Analyst [chem.] 0426
Analyst [maths] 0301
Anatom. Anzeiger 0903; 113, 169
Anatom. Gesellsch. 0903; 169
Anatom. Hefte 0911; 113, 169
Anatom. Soc. 0887
Andrews, H.C. 0683
Annalen Chem. (Pharm.) 0409; 105, 109, 150, 152
Annalen Botanik 0681

Annalen Hydrograph. etc. 1591
Annalen Pharm. See *Annalen Chem. (Pharm.)*
Annalen Physik (Chem.) 0362, 0373; 104, 109, 144, 146
Annales botan. systemat. 0697
Annales chim. analyt. 0451; 110
Annales chim. (phys.) 0393; 99, 105, 110, 145, 148, 151
Annales géogr. 0566
Annales math. pures appl. 0277
Annales mines 0455; 105, 153
Annales sci. géolog. 0488
Annales sci. nat. 0602; 158, 159, 161
Annales sci. nat.: Botan. 0689; 159
Annales sci. nat.: Zool. 0758; 161
Annales scientif. etc. Auvergne 089
Annales télégraph. 0367; 104
Annali chim. 0394; 148
Annali chim. appl. med. 0404c; 105
Annali fis. chim. mat. 0113
Annali igiene sperimentale 0931
Annali mat. pura appl. 0284; 104
Annali sci. mat. fis. 0284
Annali sci. nat. 0615
Annali sci. regno Lombardo-Veneto 091
Annals botany 0732; 161
Annals mag. nat. hist. 0614; 106, 111, 158, 164
Annals math. 0301
Annals nat. hist. 0614; 158
Annals philos. 064; 124
Annals Scott. nat hist. 0647; 111
Année biol. 0917
Annotat. zool. Japonenses 0803
Annual rep. progress chem. 0416
Antologia di Roma 96
Apoth. Verein nördl. Teutschl. 0402
Arago, F. 122
Archief wiskunde 0302
Archiv Anat. Entwick. 0872b
Archiv Anat. Physiol. [1826-32] 0869b
Archiv Anat. Physiol. [1877+] 0872b,c; 113
Archiv Anat. Physiol. wissensch. Med. 0872a; 107, 166-167
Archiv Bergbau Hüttenw. 0460a
Archiv Entwick. Organismen 0916; 113, 169
Archiv exper. Pathol. Pharmakol. 0891; 113
Archiv gesammt. Naturlehre 084
Archiv gesammt. Physiol. etc. 0889; 113, 167
Archiv Hygiene (Bakteriol.) 0925; 114
Archiv math. naturvid. 0228
Archiv Math. Phys. 0282; 104, 108
Archiv mikroskop. Anat. 0883; 113, 167
Archiv Mineral. etc. 0460b
Archiv Naturgesch. 0759; 139, 159, 163, 164, 165
Archiv pathol. Anat. Physiol. 0874; 107, 113, 167
Archiv Pharm. 0402; 105, 110
Archiv Physiol. [1795-1815] 0869a; 166
Archiv Physiol. [1877+] 0872c
Archiv wissensch. Kunde Russland 0545
Archives anat. microscop. 0921
Archives ital. biol. 0898; 113, 169
Archives néerland. sci. etc. 0197
Archives parasitol. 0936; 114
Archives physiol. normale et pathol. 0881
Archives sci. phys. nat. 0129; 107, 126, 131
Archives zool. expér. gén. 0772; 164
Archivio ital. biol. See *Archives ital. biol.*
Arkhiv biol. nauk 0913
Arnozett, C. 032
Årsber. botan. arbeten 0686
Årsber. framst. (phys.) chem. (mineral.) 0401a
Assmann 0583
Assoc. Amer. Geolog. Nat. 0470; 153
Assoc. Anat. 0923
Assoc. Franç. Avanc. Sci. 0218; 107
Astron. and astro-physics 0339b
Astron. Gesellsch. (Leipzig) 0334; 142, 143
Astron. Jahrbuch 0356a; 99, 141, 142
Astron. journ. 0331; 108, 142
Astron. Nachricht. 0329; 104, 108, 142, 143

Astron. Observatory.... *See under the name of the place or institution*
Astron. Soc. London 0328, 0330; 104, 108, 138, 142, 143
Astron. Soc. Pacific 0343
Astrophys. journ. 0339c; 108, 142
Audouin, J.V. 0602
Augsburg
  Naturhist. Verein 0626
Auk 0841; 112
Austral. Assoc. Adv. Sci. 0256
Austral. Museum 0791
Avant-coureur 177
Avicula 0852
Avicultural mag. 0851
Avicultural Soc. Brighton 0851

Baier. ... *See* Bayer. ...
Baillon, H. 0710
Balbiani, E.G. 0921
Bâle. *See* Basel
Balfour, I.B. 0732
Bamberg
  Naturf. Verein 0150
Bancroft, W.D. 0450
Banks, J. 153
Bardeleben, K. 0903
Barot, O. 0180
Bartholin, T. 95
Basel
  Naturf. Gesellsch. 099
  Soc. Physico-med. 019
Bataaf. Genootsch. Proef. Wijsbeg. 031; 94
Bataaf. Maatsch. Wetensch. 023; 94
Battaglini, G. 0288
Bauer, L.A. 0588
Baumgarten, P. 0927
Baumgartner, A. 0363
Bayer. Akad. Wissensch. 027, 063 097; 93
Bayer. Botan. Gesellsch. 0685; 159
Bayer. Botan. Gesellsch. Erforsch. heimischen Flora 0739, 0740
Bayer. Fischerei-Zeit. 0861
Bayer. Oberbergamt 0518
Behm, E. 0555
Behrens, W.J. 0679
Beilschmied, C.T. 0686
Beilstein, F. 0418; 173
Beitr. Anat. Physiol. 0880
Beitr. Biol. Pflanzen 0924; 170
Beitr. Geophysik 0587
Beitr. Natur- und Arzeneywissensch. 95
Beitr. Paläont. (Geolog.) Österreich-Ungarns 0506
Beitr. pathol. Anat. allg. Pathol. 0901; 113
Beitr. pathol. Anat. Physiol. 0901; 113, 169
Bel, K.A. 04
Belfast
  Nat. Hist. Philos. Soc. 0648
Bell, J. 0413
Ber. Fortschr. Anat. Physiol. 0877
Ber. Leist. Pflanzengeogr. etc. 0694
Ber. wissensch. Leist. Gebiete Entomol. 0808
Ber. wissensch. Leist. Naturgesch. nied. Tiere 0768
Berg- und huttenmänn. Zeit. 0471
Bergacademie Freiberg 0463; 153
Bergbaukunde 95
Bergen
  Univ. 0229
Bergmänn. Journ. 0454; 99, 152
Berlin
  Acad. Sci. 07, 017, 0356; 93, 97, 118, 120
  Akad. Wissensch. 056, 0103; 103, 107, 131
  Botan. Garten 0746
  Botan. Verein Provinz Brandenburg 0709
  Deutsch. Botan. Gesellsch. 0728; 112, 159
  Deutsch. Chem. Gesellsch. 0408, 0421; 109, 123, 151
  Deutsch. Entomol. Gesellsch. 0823, 0826
  Deutsch. Geolog. Gesellsch. 0474; 105, 110, 154
  Deutsch. Meteor. Gesellsch. 0582
  Deutsch. Physik. Gesellsch. *See* Physik. Gesell.
  Entomol. Gesellsch. 0835
  Entomol. Verein 0812
  Gesellsch. Erdkunde 0542, 0548, 0558; 111
  Gesellsch. Naturf. Freunde 0594, 0639; 93, 98, 111, 120, 157
  Physik. Gesellsch. 0362d, 0366, 0381; 109, 144, 145, 146

Physik.-tech. Reichanstalt 0380; 147
Physiol. Gesellsch. 0905; 169
Preuss. Akad. Wissensch. *See* Akad. Wissensch.
Preuss. Geolog. Landesanstalt 0505
Soc. Regia Sci. 07
Sternwarte 0356b
Univ. 151, 166
Verein Freund. Astron. 0346

Berlin. astron. Jahrbuch 0356b
Berlin. entomol. Zeitschr. 0812; 106
Berlin. Jahrbuch Pharm. 0396
Bern
Naturf. Gesellsch. 0119
Bernoulli, Jakob (I) 116
Bernoulli, Johann (III) 95
Berryat, J. 022
Berthollet, C.L. 0393
Berwickshire Naturalists' Club 0609
Berzelius, J. 0398, 0401; 139, 150
Bethune, C.J.S. 0819
Beytr. ... See also *Beitr.* ...
Beytr. chem. Annalen 0391
Biblio. anat. [Paris] 0914
Biblio. anat. [Zurich] 0918
Biblio. anc. et mod. 96
Biblio. biol. 0919
Biblio. britann. 048a; 126
Biblio. fis. Europa 95
Biblio. farm. chim. etc. 0404b
Biblio. germanique 96
Biblio. ital. 068
Biblio. math. 0309; 108
Biblio. oltremontana 96
Biblio. palaeont. 0533
Biblio. physiol. 0915
Biblio. sci. nat. 0661
Biblio. univ. Genève 048c; 126
Biblio. univ. sci. etc. 048b; 126
Bibliographia zool. 0802
Bibliotheca zool. 0788
Bierens de Haan, D. 0302
Billings, E. 0638
Biol. Anstalt Helgoland 0589
Biol. Centralbl. 0897; 113, 169
Biol. Soc. Washington 0895; 113, 169
Biol. Station Plön 0866
Biol. Zentralbl. See *Biol. Centralbl.*

Biot, J.B. 126
Black, J. 044
Board of Longit. 0354
Bode, J.E. 0356a
Böhm. Gesellsch. Wissensch. 040, 0170; 94, 97
Boll. ... See also *Bull.* ...
Boll. naturalista 0665
Bologna
Accad. Sci. 090, 0140
Sci. Artium Instit. 013; 93
Bombay
Geogr. Soc. 0544
Nat. Hist. Soc. 0667; 111
Boncompagni, B. 0291
Bonn
Naturhist. Verein 0621
Niederrhein. Gesellsch. Natur Heilkunde 0158; 103
Bonnier, G. 0735
Bonplandia 0705
Borchers, W. 0443, 0444
Bordeaux
Acad. Sci. etc. 09; 93
Soc. Anat. Physiol. 0896
Soc. Linn. 0603; 111
Soc. Sci. Phys. Nat. 0160
Borelli, G.A. 116
Born, I.E. von 033; 95
Boston
Amer. Acad. Arts Sci. 041, 0127; 94, 121
Soc. Nat. Hist. 0610, 0617; 106
Boston journ. nat. hist. 0610
Botan. bulletin 0723
Botan. Centralbl. 0725; 112, 161
Botan. Forening Kjøbenhavn 0715
Botan. Garten Berlin 0746
Botan. gazette 0723; 112
Botan. Jahrb. Systematik etc. 0726; 112, 161
Botan. Jahresber. 0721; 161
Botan. mag. [London] 0682; 159
Botan. mag. [Tokyo] 0734
Botan. mag. [Zurich] 0681
Botan. miscellany 0688a
Botan. notiser 0693; 106
Botan. repository 0683
Botan. Soc. Edinburgh 0692; 106, 159
Botan. tidsskrift 0715
Botan. Verein Gesamtthüringen 0737
Botan. Verein Provinz Brandenburg 0709
Botan. Zeit. 0698; 106, 159, 161

Botan. Zentralbl. See *Botan. Centralbl.*
Bouvier, A. 0657
Boyle, R. 116, 148
Brain 0894; 168
Brande, W.T. 124
Brandes, R. 0402, 0409a
Braune, W. 0872b
Braunschweig
Verein Naturwissensch. 0233
Braunschweig. Mag. 96
Bremen
Geogr. Gesellsch. 0561
Naturwissensch. Verein 0202
Bremiker, C. 0360
Brescia 97
Breslau
Schles. Gesellsch. Vaterländ. Cultur 061
Breslau. Samml. Natur Kunst 95
Brewster, D. 071, 083; 124
Bristol
Naturalists' Soc. 0640
Brit. Assoc. Adv. Sci. 092; 103, 107, 127
Brit. Astron. Assoc. 0345, 0348; 109, 143
Brit. Inst. Elec. Eng. 147
Brit. Meteor. Soc. 0577; 106
Brit. Museum 164
Brit. Ornithol. Club 0848; 113
Britten, J. 0713
Brno. *See* Brünn
Brongniart, A.T. 0602
Brown-Séquard, E. 0881
Brünn
Naturf. Verein 0641; 158
Brugnatelli, L. 062, 0394; 95, 148
Brunswick. *See* Braunschweig
Brussels. *See* Bruxelles
Bruxelles
Acad. Sci etc. 034, 073, 093; 94, 103
Instit. Biblio. Internat. 0915
Soc. Belge Géolog. etc. 0513
Soc. Entomol. Belge 0813
Soc. Malacol. Belg. 0855
Soc. Roy. Botan. Belg. 0712
Bryologist 0750
Buchner, J.A. 0400
Bucholz, C.F. 0392b
Budapest
Magyar Földtani Társulat 0493
Magyar Nemzeti Múzeum 0654
Magyar Tudomán. Akad. 0238, 0239, 0318
Ungar. Akad. Wissensch. 0238
Büchner, A.E. 95
Bürg, J.T. 0353
Buffon, G.L.L. de 022; 120
Buitenzorg
Jardin Botan. 0724
Bull. ... See also *Boll.* ...
Bull. astron. 0340; 109
Bull. biblio. storia sci. mat. fis. 0291
Bull. géogr. botan. 0736
Bull. pharm. See *Journ. pharm.*
Bull. sci. math. (astron.) 0296
Bull. sci. math. astron. phys. chim. 081a
Bull. sci. nat. géol. 081b
Bull. univ. sci. ind. 081, 126, 139
Bureau Longit. 0351, 0358; 153
Bureau Standards 147

Cadet, C.L. 0399a
Caen
Soc. Linn. Normandie 0601, 0637
California Acad. (Nat.) Sci. 0159
California, Univ. 0275
Cambridge, Mass.
Entomol. Club 0821
Nuttall Ornith. Club 0841
Cambridge, U.K.
Philos. Soc. 077, 0193
Univ. 168
Cambridge (Dublin) math. journ. 0281
Canadian entomologist 0819
Canadian naturalist etc. 0638
Canadian record of sci. 0638
Canstatt, C. 0873
Cape Town
South African Museum 0804
South African Philos. Soc. 0231
Carl, P. 0368
Caruel, T. 0699
Carus, J.V. 0776, 0802; 164
Časopis pěstování mat. fysik 0298
Cassel
Verein Naturkunde 0612
Catania
Accad. Gioenia Sci. Nat. 085, 0255
Cattaneo, A. 0404a,b
Cattell, J.M. 0244
Cauchy, A.L. 123

Centralbl. allg. Pathol. etc. 0909; 113, 169
Centralbl. Bakteriol. Parasitenkunde 0930; 114
Centralbl. Nahr. Genussmitt. Chem. 0447
Centralbl. Physiol. 0905; 169
Central-Zeit. Optik Mechanik 0376; 109, 146
Cesareo-Regia ... *See under the next word of the name.*
Cesaris, G.A. 0355
Česká Akad. Císarě Frant. Jos. etc. 0262
Chalon-sur-Saône
  Soc. Sci. Nat. 0656
Chamberlin, T.C. 0525
Charkov. *See* Kharkov
Charlesworth, E. 0605
Chem. abstracts 0446; 150, 152
Chem. Annalen 0391c
Chem. Archiv 0391
Chem. Centralbl. 0408; 105, 139, 150, 152
Chem. gazette 0414
Chem. Journ. 0391a
Chem. news 0419; 105
Chem. pharm. Centralbl. See *Chem. Centralbl.*
Chem. Soc. London 0412, 0435; 105, 109, 110, 124, 138, 151
Chem. Taschenb. Ärzte Chemiker Pharm. 0392b
Chem. Zentralbl. See *Chem. Centralbl.*
Chemiker Zeit. 0428; 109
Chemist 0411
Cherbourg
  Soc. Sci. Nat. 0148
Chevreul, M.E. 126
Chicago
  Univ. 0525
Christiania
  Physiograph. Forening 080
  Videnskabs Selskab 0168
Christie, W.H.M. 0337
Chun, C. 0788
Chur
  Naturf. Gesellsch. 0157
Churfürstlich ... *See under the next word of the name.*
Ciel et terre 0338
Cimento 0365; 104, 146
Cincinnati
  Soc. Nat. Hist. 0655
Circolo Mat. Palermo 0310
Clairaut, A.C. 117
Clebsch, A. 0294
Clermont-Ferrand
  Acad. Sci. etc. 089
Coast (Geod.) Survey 0546; 156
Cohn, F. 0924; 170
Colaw, J.M. 0320
Collect. acad. 022; 118
Collège de France 166
Collegio Romano, Osserv. 0332, 0578; 104, 106
Colorado Sci. Soc. 0241
Columbus
  Ohio State Acad. Sci. 0269
Comitato Geolog. Ital. 0491
Comité Ornithol. Internat. 0845
Comment. rebus in sci. nat. etc. 021; 118
Commercium litt. Norimberg. 96
Commiss. Hautes Etudes 0296
Commiss. Internat. Enseign. Math. 0325
Commiss. Ledelsen Geolog. Geogr. Undersøg. Grønland 0500
Conchol. Soc. Great Britain 0857
Conchologist 0859
Conchologist's exchange 0858
Concilium Biblio. 0533, 0802, 0915, 0918, 0919; 138, 164, 169, 176
Conferenz Internat. Erdmessung 0572
Congrès Internat. ... *See* Internat. Congress
Congrès Scientif. France 096
Congrès Soc. Savantes etc. 0274
Congressi Scientif. Ital. 0699
Congresso Scienziati Ital. 0108
Connoiss. temps 0351; 99, 141, 142, 143
Conseil (Gén.) Mines 0455; 153
Cooke, M.C. 0720
Copenhagen. *See* Kjøbenhavn.
Cornell Univ. 0388, 0450
Corresp. astron. etc. 0327
Corresp. math. phys. 0278
Corrisp. scientif. Roma 0134
Cortona 97
Cosmos 0149
Cosmos (Les mondes) 0179
Cotteswold Naturalists' F.C. 0624
Coulter, J.M. 0723
Cracovie. Cracow. *See* Kraków.
Crell, L. 0391; 99, 148, 149
Crelle, A.L. 0279

Crinon, C. 0451
Crookes, W. 0183, 0419
Croydon
Microscop. Club 0645
Cullen, W. 044
Curtis, W. 0682; 159
Curtze, M. 0304

Dalancé, J. 0351
Dames, W. 0508
Dana, E.S. 069; 125, 130
Dana, J.D. 069; 125
Daniell, J.F. 124
Danmarks Geolog. Undersøg. 0521
Dansk Geogr. Selskab 0560
Dansk Vidensk. Selskab 016, 052, 059, 082; 94
Danzig 97
Naturf. Gesellsch. 075
Darboux, G. 0296
Darmstadt
Verein Erdkunde etc. 0549
Darwin, C. 177
Daubenton, L.J.M. 022
Davis, C.H. 0361
Delage, Y. 0917
De La Rue, W. 0416b
Delesse, A. 0485
Denmark. *See* Danmark.
Dept Agric. 0584, 0790, 0824
Des Moines
Iowa Acad. Sci. 0270
Desplaces, P. 0352
Deutsch. Akad. Naturf. *See* Acad. Nat. Curios.
Deutsch. Archiv Physiol. 0869b
Deutsch. Botan. Gesellsch. 0728; 112, 159
Deutsch. botan. Monatschr. 0727
Deutsch. Bunsen-Gesellsch. 0444
Deutsch. Chem. Gesellsch. 0408, 0421; 109, 123, 151
Deutsch. Dendrolog. Gesellsch. 0742
Deutsch. Elektrochem. Gesellsch. 0444
Deutsch. Entomol. Gesellsch. 0823, 0826
Deutsch. entomol. Zeitschr. 0823, 0826
Deutsch. Gartenbau-Gesellsch. 0704
Deutsch. geogr. Blätter 0561
Deutsch. Geographentag 0563
Deutsch. Geolog. Gesellsch. 0474; 105, 110, 154
Deutsch. Gesellsch. Angew. Chem. 0439
Deutsch. Gesellsch. Kunst Wissensch. Posen 0273
Deutsch. Malakozool. Gesellsch. 0856
Deutsch. Math. Verein 0313; 108
Deutsch. Meteor. Gesellsch. 0582
Deutsch. Ornithol. Gesellsch. 0837
Deutsch. Ornithol. Verein 0836
Deutsch. Pathol. Gesellsch. 0909
Deutsch. Physik. Gesellsch. *See* Physik. Gesellsch. Berlin
Deutsch. Rundschau Geogr. Stat. 0562
Deutsch. Seewarte 0591, 0592
Deutsch. Verein Schutze Vogelwelt 0842
Deutsch. Zool. Gesellsch. 0776, 0794; 163
Devonshire Assoc. Adv. Sci. etc. 0177
Deyrolle, E. 0777
Dijon 97
Dillner, G. 0292
Döbereiner, J.W. 0396
Dohrn, F.A. 0864
Dollfus, A. 0646
Dorpat
Naturf. Gesellsch. 0101
Univ. 0271
Dove, H.W. 0364b,c; 146
Dresden
Entomol. Verein 0826
Naturwissensch. Gesellsch. 0173, 0622
Dublin
Dublin Soc. 051, 0163
Geolog. Soc. (Ireland) 0467
Irish Acad. 043, 0102; 93
Nat. Hist. Soc. 0627
Du Bois-Reymond, E. 0872a,c; 167
Duclaux, E. 0929
Ducrotay de Blainville, H.M. 030
Dumas, J.B. 0602
Du Moncel, T. 0375
Dunker, W. 0473

Echange 0782
Eckhard, C. 0880
Eclairage électr. 0375
Ecole Centrale Trav. Publ. 046

Ecole des Mines 153
Ecole Norm. Sup. 0186; 140
Ecole Polytech. 046; 121, 139, 140, 153
Ecologae geolog. Helvet. 0517
Econ. geolog. 0516; 155
Edinburgh
  Botan. Soc. 0692; 106, 159
  Geolog. Soc. 0490
  Math. Soc. 0308
  Naturalists' Field Club 0663
  Physical Soc. 0766; 144, 162
  Royal Soc. 044, 0120; 93, 97
  Scott. Geogr. Soc. 0564
  Scott. Meteor. Soc. 0576; 106
  Wernerian Nat. Hist. Soc. 0597; 154, 157
Edinburgh journ. sci. 083; 124
Edinburgh new philos. journ. 071
Edinburgh philos. journ. 071; 124
Effem. astron. Milano 0355; 99
Einstein, A. 144
Eisenmann, J.G. 0873
Elbeuf
  Soc. Etude Sci. Nat. 0664
Electrical rev. 0369
Electrician 0374
Electricien 0379; 109
Electrochimie 0448
Elektrichestvo 0378
Elektrochem. Zeitschr. 0445
Elektrotech. Verein 0377
Elektrotech. Zeitschr. 0377; 109
Elettricità 0382
Elisha Mitchell Sci. Soc. 0242
Ellis, R.L. 0281
Elwert, J.K.P. 95
Encke, J.F. 0356b
Eneström, G. 0309
Engler, A. 0726; 161
English mechanic 0194
Enseign. math. 0325; 108
Entomol. Förening Stockholm 0822
Entomol. Jahrbuch 0832
Entomol. mag. 0805
Entomol. news 0831; 112
Entomol. Soc. Canada 0819
Entomol. Soc. London 0807
Entomol. Soc. Philadelphia 0818
Entomol. Soc. Washington 0825
Entomol. tidskr. 0822
Entomol. Verein Berlin 0812
Entomol. Verein 'Iris' Dresden 0826
Entomol. Verein Stettin 0809, 0810
Entomol. Zeitschr. 0809, 0829
Entomologist 0805
Entomologist's monthly mag. 0816; 112
Entomologist's record etc. 0830
Ephem. astron. ... Coimbra 0359
Ephem. astron. merid. Vindobon. 0353; 99
Ephém. mouvem. célest. 0352
Erdmann, O.L. 0407a,b; 150
Erfurt
  Acad. Elector. Mogunt. Sci. etc. 024; 93
Erlang. gelehr. Anmerck. 96
Erlangen
  Physik. Med. Soc. 0191
Erman, G.A. 0545
Escherich, G. von 0314
Essays and obs. ... soc. Edinburgh 044
Essex Field Club 0669
Essex Instit. Salem 0165, 0642
Essex naturalist 0669
Ettingshausen, A. von 0363
Exercit. subsec. Francofurt. 96
Exner, F. 0368
Ezhegodnik geolog. mineral. Ross. 0529

Fabbri-Scarpellini, E. 0134
Faculté des Sciences ... *See under the name of the place.*
Falb, R. 0335
Faraday, M. 124, 145
Farlow, W.G. 0732
Fechner, G.T. 0364a
Fern bulletin 0743; 112
Ferrers, N.M. 0281
Ferrier, D. 0894
Férussac, A.E.d'A. 081; 126, 139
Feuille jeunes nat. 0646; 111
Fick, A. 0886; 168
Field Nat. Club Victoria 0666
Finkel, B.F. 0320
Finska Vetenskaps Soc. 0152
Firenze
  Accad. Medico-Fis. 0879
  Soc. Entomol. Ital. 0820
Fish Commiss. 0862
Fittig, R. 0418
Fitzgerald, G.F. 130
Fiz. Mat. Obshchest. Kiev. Univ. 0265
Flora 0685; 106, 159

Florence. *See* Firenze.
Flourens, P. 122, 126
Flügge, C. 0928
Fol, H. 0781
Földtani közlöny 0493
Fontenelle, B. 117
Fornasini, C. 0528
Forster, J.G.A. 95
Forster, J.R. 95
Fortschr. Physik 0366
Foster, M. 0893; 168
Fourcroy, A.F. de 0393
Francis, W. 050, 0414; 124
Frankfurt a.d. Oder
  Naturwissensch. Verein 0243
Frankfurt a.M.
  Deutsch. Malakozool. Gesellsch. 0856
  Geogr. Verein 0543
  Physik. Verein 0104; 103
  Senckenberg. Naturf. Gesellsch. 0365, 0643
  Zool. Gesellsch. 0769; 163-164
Freiberg (in Sachsen)
  Bergacademie 0463; 153
Freiburg i.B.
  Naturf. Gesellsch. 0162
Freiburg (in Schweiz). *See* Fribourg.
Fresenius, C.R. 0420
Freyberg. *See* Freiberg.
Fribourg
  Naturf. Gesellsch. 0659
  Soc. Sci. Nat. 0659
Friedel, C. 0453
Froriep, L.F. von 076
Froriep, R. 076
Funk, C.B. 95
Fusinieri, A. 091

Gaea 0190
Galleria Minerva 96
Gallois, J. 01
Garden 0718
Gardener's chronicle 0695; 112
Garnett, T. 125
Garnier, J.G. 0278
Gartenflora 0704; 112
Gauss, C.F. 0586; 140
Gazz. chim. ital. 0424
Gefiederte Welt 0839
Gegenbauer, C. 0892; 168
Gegenbauer, L. 0314
Gehlen, A.F. 0396, 0397b,c, 0400; 149
Geiger, P.L. 0403b,c, 0409a
Genève
  Conserv. et Jardin Botan. 0748
  Herbier Boissier (Chambésy) 0744
  Musée Hist. Nat. 0781
  Soc. Phys. Hist. Nat. 078
Genoa. *See* Genova.
Genova
  Soc. Ligustica Sci. Nat. 0261
  Univ. 0795
Gentlemen's mag. 96
Gentlemen's math. companion 96
Geogr. Gesellsch. Bremen 0561
Geogr. Gesellsch. Wien 0552, 0571; 105
Geogr. Jahrbuch 0555
Geogr. journ. 0550; 110
Geogr. Soc. London 0541, 0550; 105
Geogr. tidsskr. 0560
Geogr. Verein Frankfurt 0543
Geogr. Zeitschr. 0569; 111
Geolog. Commiss. Cape of Good Hope 0531
Geolog. Förening. Stockholm 0495
Geolog. Komiteta St. Peterburg 0510, 0511
Geolog. Landesanstalt Elsass-Lothringen 0515
Geolog. mag. 0482b; 110
Geolog. Mineral. Museum Leiden 0507
Geolog. (Nat. Hist.) Survey Canada 0512
Geolog. Nat. Hist. Survey Minnesota 0745
Geolog. Polytech. Soc. West Riding Yorkshire 0468
Geolog. record 0498
Geolog. Reichanstalt Wien 0475, 0481; 105, 110
Geolog. Soc. America 0522; 110, 154, 155
Geolog. Soc. Cornwall 0459; 154
Geolog. Soc. Dublin 0467
Geolog. Soc. Glasgow 0484
Geolog. Soc. Ireland 0467
Geolog. Soc. London 0457, 0462; 105, 110, 138, 153
Geolog. Soc. South Africa 0532
Geolog. Survey England 0472
Geolog. Survey Great Britain 0472
Geolog. Survey India 0480, 0487
Geolog. Survey Indiana 0489
Geolog. Survey New South Wales 0520

Geolog. Survey United Kingdom
0472, 0535
Geolog. Survey United States
0504; 110
Geologist 0482a
Geologists' Assoc. 0486; 154
Gergonne, J.D. 0277
Gerland, G. 0587
Germer, E.F. 0810
Gesellsch. Beförd. gesammt.
Naturwissensch. Marburg 0198
Gesellsch. Deutsch. Naturf.
Ärzte 079; 104, 107, 127
Gesellsch. Erdkunde Berlin
0542, 0548, 0558; 111
Gesellsch. Morphol. Physiol.
München 0902
Gesellsch. Naturf. Freunde Berlin
0594, 0639; 93, 98, 111, 120,
157
Gesellsch. Wissensch. Göttingen
015, 0105, 0124
Giessen 150
Oberhess. Gesellsch. Natur-
und Heilkunde 0133
Gilbert, L.W. 0362b; 144
Giorn. arcadico sci. etc. 072
Giorn. botan. ital. 0699
Giorn. encicloped. Italia 96
Giorn. farm. chim. etc. 0404a
Giorn. fis. chim. storia nat. 062
Giorn. Italia 96
Giorn. lett. Italia 96
Giorn. lett. Italia [ed. Zeno]
08; 93, 98
Giorn. mat. 0288; 104
Giorn. Roma 96
Giorno 0382
Glasgow
Geolog. Soc. 0484
Philos. Soc. 0114
Globus 0553; 111
Gloucester
Cotteswold Naturalists' Field
Club 0624
Gmelin, C.G. 0401b
Godin, L. 0351
Göttingen
Gesellsch. Wissensch. 015,
0105, 0124
Magnet. Verein 0586
Soc. Regia Sci. 020; 93
Götting. (gelehr.) Anzeigen 015
Götting. Mag. Wissensch. Lit. 95
Götting. Zeit. gelehr. Sachen
015; 93, 98
Göttling, J.F.A. 0392a
Gornuy zhurnal 0461
Gould, B.A. 0331; 142
Graz
Naturwissensch. Verein 0181
Gregory, D.F. 0281
Greifswald
Naturwissensch. Verein 0208
Gren, F.A.C. 0362a; 144
Grevillea 0720
Grew, N. 116
Grisebach, A. 0694; 159
Groth, P. 0501
Grunert, J.A. 0282
Guccia, G.B. 0310
Günther, A.C.L.G. 0771
Günther, S. 0304
Guérin-Méneville, F.E. 0755,
0760, 0764; 161
Güstrow
Verein Freunde Naturgesch.
Mecklenburg 0625
Guide du naturaliste 0657
Guyton de Morveau, L.B. 0393

Haarlem
Bataaf. Maatsch. Wetensch.
023; 94
Holland. Maatsch. Wetensch.
023, 0197
Musée Teyler 0196
Teyler's Stichting 0196
Hänle, G.F. 0403a
Hale, G.E. 0339b,c
Halle
Deutsch. Math. Verein 0313; 108
Deutsch. Verein Schutze
Vogelwelt 0842
Naturf. Gesellsch. 0153
Naturwissensch. Verein 0154
Halle. wöchentl. Anzeigen 96
Haller, A. 015
Hamburg
Deutsch. Seewarte 0591, 0592
Math. Gesellsch. 0299
Naturwissensch. Verein 0226
Wissensch. Anstalt 0245
Hamburg. Mag. 96
Hanau. Mag. 96
Hann, J.F. von 0582
Hannöver. gelehr. Anzeigen 96
Hannöver. Mag. 96
Hannover
Naturhist. Gesellsch. 0629

Hardwicke's sci. gossip 0192
Harlem. *See* Haarlem.
Hartmann, C. 0471
Harvard College
Astron. Observ. 0333; 104
Museum Compar. Zool. 0770
Hébert, E. 0488
Hedwigia 0703
Heidelberg
Naturhist. Med. Verein 0166
Hell, M. 0353
Helsingfors
Finska Vetenskaps Soc. 0152
Soc. Fauna Flora Fennica 0650, 0653
Soc. Sci. Fennica 0109
Helsinki. *See* Helsingfors.
Hendricks, J.E. 0301
Henle, J. 0877
Henneguy, L.F. 0921
Herbier Boissier 0744
Hermann, L. 0877
Herrick, C.L. 0910
Hertfords. Nat. Hist. Soc. 0658
Hertwig, O. 0883
Hessische Beitr. 96
Hevelius, J. 116
Higgs, W. 0369
Himmel und Erde 0257
Hindenburg, K.F. 95
Hingenau, O. von 0479
Hirsch, A. 0888
His, W. 0872b
Hisinger, W. 0398
Hist. ouvrages savans 96
Hobart
Royal Soc. Tasmania 0138
Höpfner, J.G.A. 95
Hoff, K.E.A. von 125
Hoffmann, C.A.S. 0454; 153
Hoffmann, J.C.V. 0295
Hofmann, A.W. 0416b; 151
Hofmann, F. 0877
Holland. Maatsch. Wetensch. 023, 0197
Holmgren, F. 0907; 169
Holzbecher 014b
Hooker, J.D. 0682, 0691
Hooker, W.J. 0682, 0688, 0691; 159
Hooker's journ. botany 0688d
Hooreman, C. 0338
Hoppe-Seyler, F. 0427
Horticult. Soc. London 0684, 0700
Hospitalier, E. 0387
Hübner, H. 0418
Hübner, L. 95
Humphrey, G.M. 0887
Husnot, T. 0722
Hydrograph. Amt Pola 0590
Hydrograph. Mitth. 0591

Ibis 0838
Icones plantarum 0691
Illustr. Wochenschr. Entom. 0835
Illustr. Zeitschr. Entom. 0835
Imperial ... *See under the next word of the name.*
Indiana Acad. Sci. 0264
Indust. électr. 0387
Innsbruck
Naturwissensch. Med. Verein 0213
Insectenbörse 0827
Instit. Biblio. Internat. Bruxell. 0915
Instit. Eksper. Med. 0913
Instit. Elec. Eng. 0389
Instit. National de France 047, 057; 121, 122, 153
Instit. Pasteur 0929; 114
Institut: Journ. acad. etc. 095
Internat. cat. sci. lit. 101, 138, 155, 171, 175
Internat. Conf. (Plant Breeding) Hybridisation 0752
Internat. Congress Botany 0714
Internat. Congress Geogr. 0557; 111
Internat. Congress Math. 0324
Internat. Congress Physiol. 0908
Internat. Congress Pure and Appl. Chem. 0442
Internat. Entomologenverein 0828, 0829
Internat. Geogr. Congress. *See* Internat. Congress Geogr.
Internat. Geolog. Congress 0502; 110
Internat. Meteor. Conference 0581
Internat. Monatschr. Anat. (Histol.) (Physiol.) 0900a
Internat. Ornithol. Congress 0844
Internat. Zool. Congress 0789
Introd. observ. phys. etc. 030
Iowa Acad. Sci. 0270
Iowa Geolog. Survey 0523
Irish Acad. 043, 0102; 93

Irish naturalist 0672; 111
Isis 0754; 127, 162-163
Istituto Lombardo Sci. etc. 0111, 0185
Istituto Veneto Sci. etc. 0112

Jahrbuch Astron. Geophys. 0344; 144
Jahrbuch Berg- und Hüttenmann 0463
Jahrbuch Berg- und Hüttenwesen 0463
Jahrbuch Chem. 0441
Jahrbuch Elektrochem. 0443
Jahrbuch Fortschr. Math. 0293
Jahrbuch Mineral. etc. 0456b; 105, 110, 153
Jahrbücher wissensch. Botanik 0708; 161, 163
Jahresber. Arbeiten physiol. Botanik 0690b
Jahresber. Fortschr. Anat. Entwickl. 0877
Jahresber. Fortschr. Anat. Physiol. 0877
Jahresber. Fortschr. Biol. 0873
Jahresber. ... Fortschr. Botanik 0686
Jahresber. Fortschr. ... Chem. etc. 0416; 152
Jahres-Ber. Fortschr. Chem. (Mineral.) 0401b; 139
Jahresber. Fortschr. Lehre Gährungs-Organismen 0932
Jahresber. Fortschr. Lehre path. Mikroorganismen etc. 0927
Jahresber. Fortschr. Leist. Gebiete Hygiene 0926
Jahresber. Fortschr. Physiol. 0877
Jahres Ber. Fortschr. physisch. Wissensch. 0401b
Jahresber. Fortschr. Tierchem. 0423
Jahresber. Leist. Fortschr. Anat. Physiol. 0888
Jahresber. Leist. physiol. Wissensch. 0873
Jahresber. Resultate ... physiol. Botanik 0690a
Jameson, R. 071, 0597
Jardin Botan. Buitenzorg. 0724
Jardine, W. 0614
Järnkontor. *See* Jernkontor.
Jeaurat, E.S. 0351
Jednota Česk. Mat. Fysik 0298
Jena 97
  Med. Naturwissensch. Gesellsch. 0187
Jena. gelehr. Zeit. 96
Jena. Zeitschr. (Med.) Naturwissensch. 0187
Jernkontor 0458
Johns Hopkins Univ. 0234, 0305; 168
Jones, T.R. 0482b
Jornal sci. math. phys. nat. 0203
Journ. anat. (physiol.) 0887; 113, 168
Journ. anat. physiol. normal. et pathol. etc. 0882
Journ. botanique [Paris] 0733
Journ. botanique [Kjøbenhavn] 0715
Journ. botany [1834-42] 0688b
Journ. botany [1863+] 0713; 106, 112, 160
Journ. Chem. Physik 0397c,d; 145, 149
Journ. chim. méd. 0405; 105, 148
Journ. comp. neurol. 0910; 169
Journ. conchol. 0857
Journ. conchyliol. 0854
Journ. encycloped. 96
Journ. Genève 97
Journ. geolog. 0525; 110, 155
Journ. hist. nat. 177
Journ. Liebhaber Steinreichs Conchyliol. 95
Journ. litt. 96
Journ. litt. Allemagne 96
Journ. litt. La Haye 96
Journ. litt. sci. arts 06
Journ. malacol. 0859
Journ. math. pures appl. 0280; 104
Journ. mines 0455; 153
Journ. morphol. 0906; 169
Journ. mycology 0729
Journ. nat. philos. etc. 049; 123
Journ. Ornithologie 0837
Journ. Paris 96
Journ. pathol. bacteriol. 0933; 114
Journ. pharm. 0399b; 105, 148
Journ. Pharm. Aerzte Apoth. 0395
Journ. pharm. chim. 0399c
Journ. physical chem. 0450
Journ. Physik. See *Annalen Physik.*
Journ. physiol. 0893; 113, 168, 169

Journ. physiol. expér. 0870; 166
Journ. physiol. homme animaux 0881; 107
Journ. physiol. pathol. gén. 0881
Journ. physique 030; 148
Journ. physique théor. appl. 0370; 109, 146
Journ. prakt. Chem. 0407b; 105, 150
Journ. reine ang. Math. 0279; 104, 108
Journ. savants. See *Journ. sçavans.*
Journ. sçavans 01; 93, 98, 115, 116, 126, 139
Journ. sci. 0183
Journ. sci. arts. See *Quart. journ. lit., sci. and arts*
Journ. sciences beaux arts 06
Journ. tech. ökon. Chem. 0407a
Journ. Trévoux 06
Jurjev. *See* Dorpat.
Just, L. 0721
Justus Liebig's Annalen Chem. See *Annalen Chem.*
Justus Perthes' Geogr. Anstalt 0551; 105

Kästner, A.E. 014b
Kaiserlich ... *See under the next word of the name.*
Kalendar sächsisch. Berg- und Hüttenmann 0463
Kansas Acad. Sci. 0204
Kansas Nat. Hist. Soc. 0204
Kansas, Univ. 0267
Karlsruhe
  Naturwissensch. Verein 0178
Karsten, C.J.B. 0460; 153
Karsten, D.L.G. 120
Kassel. *See* Cassel.
Kastner, C.(or K.) W.G. 084, 0396
Kayser, E. 0508
Kazan
  Obshchest. Estest. 0215
Kekulé, A. 0418
Kelvin, *Lord.* *See* Thomson, W.
Kemiska notiser 0438
Kemistssamfundet, Stockholm 0438
Kew
  Botanic Gardens 0682, 0731
Kharkov
  Mat. Obshchest. 0306
  Univ. 0263
Kiel
  Naturwissensch. Verein 0221
Kiev
  Fiz. Mat. Obshchest. 0265
  Obshchest. Estest. 0210
Kjøbenhavn
  Botan. Forening 0715
  Dansk Geogr. Selskab 0560
  Dansk Vidensk. Selskab 016, 052, 059, 082; 94
  Naturhist. Forening 0628; 111
Klagenfurt
  Naturhist. Museum 0765, 0793
Klaproth, M.H. 95, 120
Klebs, E. 0891
Klein, H.J. 0190
Klein, J. 0344
Klein's Jahrbuch Astron. Geophys. 0344
Klipstein, P.E. 95
Klobükow 0445
Knaggs, H.G. 0816
Kneucker, A. 0747
Knowledge 0235
Koch, A. 0932
Koch, R. 0928; 170
Köhler, A.W. 0454; 153
Kölliker, A. 0763; 163
Königlich ... *See under the next word of the name.*
Königsberg
  Physik. Ökon. Gesellsch. 0172
  Preuss. Botan. Verein 0711
Kolbe, H. 0407b; 176
Kommiss. Wissensch. Untersuch. Deutsch. Meere 0589
Kopp, H. 0416a,b; 150
Kosmos 0227
Krahmann, M. 0526
Kraków
  Akad. Umiej. 0224, 0259; 107
  Towarz. Nauk 0199
  Towarz. Rybackiego 0863
Krause, G. 0428
Krause, W. 0900a
Kristiania. *See* Christiania.
Krit. Zeitschr. Chem. etc. 0418
Krøyer, H. 0611
Krüss, G. 0440
Kundmann, J.C. 95
Kunstrechnungslieb. Soc. Hamburg 0299
Kurfürstlich ... *See under the next word of the name.*

Lacaille, N.L. de 0352
Lacaze-Duthiers, H. de 0772
Ladies' diary 96
Lalande, J.J. de 0351, 0352; 117
Lambert, J.H. 95
Lamétherie, J.C. de 030; 148
Lankester, E. 0675
Lapparent, A. 0485
La Rive, A.A. de 048c, 0129; 126
Lausanne
  Soc. Vaud. Sci. Nat. 0618; 111
Lausitz. Mag. 96
Lavoisier, A.L. 0393; 148, 177
Leeds
  Astron. Soc. 0349
  Yorkshire Geolog. Soc. 0468
Leeuwenhoek, A. van 116, 170
Le Fèvre, J. 0351
Leibniz, G.W. 116, 118
Leicester
  Lit. Philos. Soc. 0100
Leiden
  Geolog. Mineral. Museum 0507
  Museum Nat. Hist. 0778
  Nederland. Entomol. Vereen. 0811
  Univ. 0383; 146
Leimbach, G. 0727
Leipzig 116, 127, 150
  Astron. Gesellsch. 0334; 142, 143
  Naturf. Gesellsch. 0225
  Sächsisch. Gesellsch. Wissensch. 0126, 0139
  Univ. 0885; 168
Leipzig. Mag. Naturkunde Math. Oekon. 95
Leipzig. Mag. reine angew. Math. 95
Leoben
  Montan-Lehranstalt 0477
Leonhard, K.C. von 0456; 153
Leopold-Carolin. Akad. Naturf. *See* Acad. Nat. Curios.
Leske, N.G. 95
Leuckart, R. 0768, 0788
Leyden. *See* Leiden.
Lichtenberg, G.C. 95
Lichtenberg, L.C. 035
Lick Observatory 0342
Lie, S. 0228
Liebig, J. 0403c, 0409, 0416a; 150, 152, 166, 176
Liège
  Soc. Sci. 0117
Lieutaud, J. 0351
Lille
  Soc. Amateurs Sci. Arts 053
  Soc. Géolog. Nord 0492
  Soc. Sci. 053
Lindblom, A.E. 0693
Lindley, J. 0695
Link, H.F. 0690
Linnaea 0687; 159
Linnaea entomol. 0810
Linnaean ... *See also* Linnean ...
Linnaean fern bulletin 0743
Linnaeus, C. 120
Linnean Soc. London 0595, 0613, 0707, 0767; 138, 157, 158, 159, 160, 162
Linnean Soc. New South Wales 0652; 111
Liouville, J. 0280; 140
Lisboa
  Acad. Sci. 0203
Listy chem. 0425
Lit. Helvet. 96
Liverpool
  Biol. Soc. 0904
  Geolog. Soc. 0483
Lockyer, J.N. 0206
London
  Astron. Soc. 0328, 0330; 104, 108, 138, 142, 143
  Board of Longitude 0354
  Brit. Astron. Assoc. 0345, 0348; 109, 143
  Brit. Meteor. Soc. 0577; 106
  Chem. Soc. 0412, 0435; 105, 109, 110, 124, 138, 151
  Entomol. Soc. 0807
  Geogr. Soc. 0541, 0550; 105
  Geolog. Soc. 0457, 0462; 105, 110, 138, 153
  Geologists' Assoc. 0486; 154
  Horticult. Soc. 0684, 0700
  Linnean Soc. *See main entry.*
  Malacol. Soc. 0860
  Math. Soc. 0289; 108, 138
  Meteor Soc. 0577
  Microscop. Soc. 0674
  Museum Econ. Geol. 0472
  Neurolog. Soc. 0894
  Physical Soc. 0372, 0389; 146, 147
  Quekett Microscop. Club 0676; 112
  Royal Instit. 067, 0145; 124
  Royal Soc. *See main entry.*

London (cont'd)
Soc. Chem. Ind. 0433; 109, 152
Soc. Public Analysts 0426
Zool. Soc. 0756, 0757, 0771; 106, 112, 138, 162, 164
London Edinburgh (Dublin) philos. mag. See *Philos. mag.*
London journ. botany 0688c
Lorgna, A.M. 037; 121
Lotos 0631
Loudon, J.C. 0605
Ludwig, C. 0885; 168
Ludwig, C.G. 021
Lumière électr. 0375; 109
Lund
Univ. 0188
Lundh, G.F. 080
Luxford, G. 0696
Lwów
Polsk. Towarz. Przyrod. 0227
Lyceum Nat. Hist. New York 0599
Lyon
Acad. Sci. etc. 0121
Soc. Agric. etc. 058
Soc. Botan. 0717
Soc. Linn. 0604; 106

Mackie, S.J. 0482a
Madison
Wisconsin Acad. Sci. etc. 0209
Madrid
Acad. Ciencias 0142
Mag. Apoth. Materialisten Chemisten 95
Mag. Botanik 0681
Mag. nat. hist. 0605; 158
Mag. nat. hist. and Naturalist 0632a
Mag. Naturkunde Helvetiens 95
Mag. naturvidenskab. 080
Mag. neuest. Erfahr. ... Pharm. 0403a
Mag. Neuest. aus Physik und naturgesch. 035; 93, 98, 121
Mag. neuest. Zustand Naturkunde 035
Mag. Pharm. etc. 0403b
Mag. Pharm. Exp. Kritik 0403c
Mag. zool. 0755
Magendie, F. 0870; 166
Magnet. Verein 0586
Magyar chem. folyóirat 0449
Magyar Földtani Társulat 0493
Magyar Nemzeti Múzeum 0654
Magyar Orvosok Termész. Nagy. 0110
Magyar Termész. Társulat 0207, 0239
Magyar Tudomán. Akad. 0238, 0239, 0318
Mainz. Akad. nutzl. Wissensch. zu Erfurt 024
Majocchi, G.A. 0113
Malacol. Soc. London 0860
Malakazool. Blätter 0853
Malpighia 0730
Maly, R. 0423
Manchester
Geolog. Soc. 0469
Lit. Philos. Soc. 039, 0167; 93
Microscop. Soc. 0678
Mandeix, A. 0386
Mannheim
Acad. Elector. Sci. etc. 028; 93
Mansion, P. 0303
Maraldi, G.D. 0351
Marburg
Gesellsch. Beförd. gesammt. Naturwissensch. 0198
Marignac, J.C. 0129
Marine Biol. Assoc. U.K. 0865
Marine Biol. Lab. Woods Hole 0868
Marpmann, G. 0680
Marseul, S.A. de 0817
Martin, H.N. 168
Maryland Geolog. Survey 0536
Maskelyne, N. 0354
Mat. sbornik 0290
Math. Annalen 0294; 108
Math. Gesellsch. Hamburg 0299
Math. naturwissensch. Ber. aus Ungarn 0239
Math. naturwissensch. Mitt. 0247
Math. Naturwissensch. Verein Württemberg 0247
Math. phys. lapok 0318; 108
Math. tidsskr. 0286
Mathésis 0303
Matteucci, C. 0365a,b; 146
Maunder, E.W. 0345
Maurice, F.G. 048a
Méchain, P.F.A. 0351
Meckel, J.F. 0869b; 166
Med. Naturwissensch. Gesellsch. Jena 0187
Meddl. Grønland 0500
Meissner, G. 0877
Mélanges philos. math. Turin 025; 94
Melbourne

Field Naturalists' Club
Victoria 0666
Royal Soc. Victoria 0161
Mém. pour servir hist. sci. etc. See *Mém. Trévoux.*
Mém. Trévoux 06; 93, 98, 117
Mencke, O. 04
Mendel, J.G. 158
Menke, K.T. 0853
Mercure de France 96
Merritt, E. 0388
Messenger math. 0287; 104, 108
Meteor. mag. 0579; 111
Meteor. Soc. London 0577
Meteor. vestnik 0585
Meteor. Zeitschr. 0582; 111
Metz
Soc. Hist. Nat. 0619
Meyen, F.J.F. 0690; 159
Meyer, H. von 0473
Meyer, R. 0441
Michigan Acad. Sci. 0272
Microscop. Soc. London 0674
Milano
Istituto Lombardo Sci. etc. 0111, 0185
Osserv. Brera 0355
Soc. Geolog. 0636
Soc. Ital. Sci. Nat. 0636
Milne-Edwards, A. 0488
Mineral. Briefwechsel 95
Mineral. mag. journ. 0499
Mineral. Mitth. 0494
Mineral. Obshchest. 0465
Mineral. petrograph. Mitth. 0494; 110
Mineral. Soc. Great Britain 0499
Minet, A. 0448
Minnesota botan. studies 0745
Misc. Berolin. 07; 118
Misc. curios. med. phys. 03; 116
Misc. Lipsiensia 96
Misc. philos. math. soc. privat. Taurin. See *Mélanges philos. math. Turin.*
Misc. phys. med. math. 95
Misc. Taurin. See *Mélanges philos. math. Turin.*
Mittag-Leffler, G. 0307
Mitth. Fischereiwesen 0861
Mitth. Gebiete Seewesens 0590
Modena
Soc. Naturalisti 0200
Mohl, H. 0698
Moigno, F. 0149, 0179
Mojsisovics, E. von 0506
Moleschott, J. 0878
Monatshefte Chem. 0431; 109
Monatshefte Math. (Physik) 0314; 108
Monatl. Corresp. Beförd. Erd- und Himmelskunde 0326
Monatsschr. Kakteenkunde 0738
Monde des plantes 0736
Mondes 0179; 103
Monfort, B.R. de 0149
Monge, G. 0393
Mongez, J.A. 030
Monit. scient. de Quesneville 0410b; 105
Monit. scient. du chimiste etc. 0410b
Monit. zool. ital. 0792
Montan-Lehranstalt Leoben etc. 0477
Monthly internat. journ. anat. etc. 0900b
Monthly weather rev. 0584; 111
Montpellier
Acad. Sci. etc. 0131; 174
Montreal
Nat. Hist. Soc. 0638
Royal Soc. Canada 0237
Morphol. Arbeiten 0912; 169
Morphol. Jahrbuch 0892
Morris, B.R. 0632
Moscow. *See* Moskva.
Moser, L. 0364b,c
Moskau. *See* Moskva.
Moskov. ... *See* Moskva.
Moskva
Mat. Obshchest. 0290
Soc. Imp. Naturalistes 0596, 0606; 106, 111
Mosso, A. 0898; 169
Müller, F. 0293
Müller, J. 0872a; 166
München
Bayr. Akad. Wissensch. 027, 063, 097; 93
Bayr. Botan. Gesellsch. Erforsch. Heimischen Flora 0739, 0740
Bayr. Oberbergamt 0518
Gesellsch. Morphol. Physiol. 0902
Mulder, G.J. 0415
Munich. *See* München.
Musée Hist. Nat. Genève 0781
Musée Teyler 0196
Museum Compar. Zool. Harvard 0770
Museum Econ. Geol. London 0472

Museum Nat. Hist. Leiden 0778
Muséum Hist. Nat. Paris 0753, 0800; 112, 161, 164
Muséum Hist. Nat. Pays-Bas 0778

Nägeli, C.W. von 163
Nancy
  Acad. Stanislas 054
  Soc. Sci. etc. 054, 0222
Nantes
  Soc. Sci. Nat. Ouest 0670
Naples. *See* Napoli.
Napoli
  Accad. Sci. 070, 0115, 0174, 0175; 103
  Soc. Naturalisti 0668
  Soc. Reale Borbonica 070
  Zool. Station 0864; 165
Nassau. Verein Naturkunde 0119
Nat. hist. rev. 0634
Nat. Hist. Soc. Dublin 0627
Nat. Hist. Soc. Montreal 0638
Nat. Hist. Soc. Northumberland etc. 0608
Nat. Sci. Assoc. Staten Is. 0246
National Acad. Sci. 0182
National geogr. mag. 0565
National Geogr. Soc. 0565
National Microscop. Congr. 0677
National Museum 0780
National Phys. Lab. 147
Natur 0147; 104, 107
Natur und Haus 0671; 111
Naturae novitates 0660
Naturalist 0632; 111
Naturaliste 0777; 112
Nature [London] 0206; 107, 130, 131
Nature [Paris] 0220; 107
Naturen 0229
Naturf. Gesellsch. Bamberg 0150
Naturf. Gesellsch. Basel 099
Naturf. Gesellsch. Bern 0116
Naturf. Gesellsch. Danzig 075
Naturf. Gesellsch. Freiburg 0659
Naturf. Gesellsch. Freiburg i.B. 0162
Naturf. Gesellsch. Graubünden 0157
Naturf. Gesellsch. Halle 0153
Naturf. Gesellsch. Leipzig 0225
Naturf. Gesellsch. Rostock 0249
Naturf. Gesellsch. Univ. Jurjev 0151
Naturf. Gesellsch. Zürich 0130
Naturf. Verein Bamberg 0150
Naturf. Verein Brünn 0641; 158
Naturf. Verein Riga 0123
Naturforscher 0593; 93, 98, 120, 157
Naturhist. Forening Kjøbenhavn 0628; 111
Naturhist. Gesellsch. Hannover 0629
Naturhist. Hofmuseum Wien 0783
Naturhist. Med. Verein Heidelberg 0166
Naturhist. Museum Kärnten 0765, 0793
Naturhist. tidsskr. 0611
Naturhist. Verein Augsburg 0626
Naturhist. Verein 'Lotos' Prag 0631
Naturhist. Verein Preussisch. Rheinlande 0621
Naturwissensch. Gesellsch. 'Isis' Dresden 0173, 0622
Naturwissensch. Med. Verein Innsbruck 0213
Naturwissensch. Rundshau 0250
Naturwissensch. Verein Bremen 0202
Naturwissensch. Verein Frankfurt a.d. Oder 0243
Naturwissensch. Verein Halle 0154
Naturwissensch. Verein Hamburg 0226
Naturwissensch. Verein Karlsruhe 0178
Naturwissensch. Verein Neu-Vorpommern Rügen 0208
Naturwissensch. Verein Regensburg 0623
Naturwissensch. Verein Schleswig-Holstein 0221
Naturwissensch. Verein Schwaben Neuburg 0626
Naturwissensch. Verein Steiermark 0181
Naturwissensch. Wochenschr. 0251; 107
Natuurkund. tijdsch. Nederland Indië 0630; 106
Naumannia 0836
Naunyn, B. 0891
Naut. almanac 0354; 141
Naut. Jahrbuch 0360
Nautilus 0858; 113
Neapel. *See* Napoli.
Nebraska Univ. Zool. Lab. 0799

Nederland. Aardrijsk. Genootsch. 0559
Nederland. Botan. Vereen. 0701
Nederland. Dierkund. Vereen. 0773
Nederland. Entomol. Vereen. 0811
Nederland. kruidkund. archief 0701
Nederland. Nat. Geneesk. Congress 0254
Nernst, W. 0443
Neu ... See under the next word of the title.
Neuchâtel
  Soc. Sci. Nat. 0620
Neuest. ... See under the next word of the title except in the following case.
Neuest. Entdeck. Chem. 0391b
Neumann, C. 0294
Neumayr, M. 0506
Neurolog. Soc. London 0894
Neurolog. Zentralbl. 0899; 169
New ... *Newest* ... See under the next word of the title.
New England Botan. Club 0751
New York
  Acad. Sci. 0230, 0599
  Amer. Geogr. (Stat.) Soc. 0547
  Amer. Museum Nat. Hist. 0779
  Botan. Club 0716
  Entomol. Soc. 0833
  Lyceum Nat. Hist. 0599
  Math. Soc. 0317
  Torrey Botan. Club 0716; 112
New York State Museum 0762, 0786
New Zealand Instit. 0205
Newcastle-upon-Tyne
  Nat. Hist. Soc. 0608
Newcomb, S. 0361
Newman, E. 0696, 0761, 0805
Nichols, E.L. 0388
Nicholson, W. 049; 123-124
Nicolas, A. 0914
Niederrhein. Gesellsch. Natur- und Heilkunde 0158; 103
Nieuw ... See under the next word of the title.
Niewenglowski, B. 0315
Nîmes
  Soc. Etude Sci. Nat. 0649
Norddeutsch. Seewarte 0592
Norfolk ... Naturalists' Soc. 0644
North American fauna 0790
North Carolina, Univ. 0242
Northamptonsh. Nat. Hist Soc. 0662
Norwich
  Naturalists' Soc. 0644
Notizen Gebiete Natur- und Heilkunde 076
Nouveau ... *Nouvelle* ... See under the next word of the title except in the following cases.
Nouv. ann. math. 0283; 104, 108, 140
Nouv. biblio. 96
Nouv. corresp. math. 0303
Nouv. républ. lettres 96
Nova ... See under the next word of the title.
Novitates zool. 0797
Novoross. Obshchest. Estest. 0219
Noyes, A.A. 0446
Nuova ... *Nuovi* ... *Nuovo* ... See under the next word of the title.
Nuttall Ornithol. Club 0841
Nya ... *Nyt* ... See under the next word of the title.

Oberhess. Gesellsch. Natur- und Heilkunde 0133
Observ. sur la physique 030
Observatory 0337; 109, 144
Observatory ... *See under the name of the place or institution.*
Obshchest. Estest. Kazan Univ. 0215
Odessa
  Novoross. Obshchest. Estest. 0219
Österreich. botan. Wochenblatt 0702; 161
Österreich. botan. Zeitschr. 0702; 106
Österreich. Gesellsch. Meteor. 0580
Österreich. Zeitschr. Berg- und Hüttenwesen 0479
Ohio State Acad. Sci. 0269
Ohrtmann, C. 0293
Oken, L. 0754; 127, 162-163
Okólnik rybacki 0863
Olbers, H.W.M. 95
Oldenburg, H. 02; 115
Olivier, L. 0260
Onnes, H.K. 0383
Opusc. scelti sci. arti. See *Scelta opusc. interess.*
Ornis 0845
Ornithol. Jahrbuch 0846
Ornithol. Monatsber. 0850; 112
Ornithol. Monatsschr. 0842; 113

Ornithol. Verein Stettin 0840
Oslo. *See* Christiania.
Osserv. Brera 0355
Osserv. lett. 96
Ostwald, W. 0437, 0444
Oxford Cambridge Dublin messenger math. 0287

Packard, A.S. 0642
Padova
  Accad. Sci. 042; 94, 121
Padua. *See* Padova.
Palaeont. Abhandl. 0508
Palaeont. ital. 0534
Palaeontographica 0473
Palermo
  Circ. Mat. 0310
Panebianco, R. 0514
Papin, D. 116
Paris
  Acad. Sci. *See main entry.*
  Agence des Mines 0455; 153
  Bureau des Longit. 0351, 0358; 153
  Conseil (Gén.) des Mines 0455; 153
  Ecole Norm. Sup. 0186; 140
  Ecole Polytech. 046; 121, 139, 140, 153
  Institut de France 047, 057; 121, 122, 153
  Institut Pasteur 0929; 114
  Muséum Hist. Nat. 0753, 0800; 112, 161, 164
  Observatoire 0340
  Soc. Anat. 0871; 113, 166
  Soc. Arcueil 060; 122
  Soc. Astron. France 0341; 109, 143
  Soc. Biol. 0876; 107, 113, 166, 168, 169
  Soc. Botan. France 0706; 106, 112, 160
  Soc. Chim. 0417; 105, 109, 151
  Soc. Chim. Méd. 0405
  Soc. Entomol. France 0806, 0834; 106, 112
  Soc. Franç. Phys. 0371; 146
  Soc. Géogr. 0540; 105, 111
  Soc. Géolog. France 0464, 0466; 105, 110, 153
  Soc. Hist. Nat. 0600; 158
  Soc. Math. 0297; 108
  Soc. Météor. France 0575
  Soc. Pharm. 0399b
  Soc. Philomat. 045; 104, 122
  Soc. Zool. France 0774, 0787; 163
Parmentier, A.A. 0399a
Pasteur, L. 0186; 170
Paternò, E. 0424
Pathol. Soc. Gr. Britain 0933
Pavia
  Univ. 0719
Payne, W.W. 0339a,b, 0350
Peabody Acad. Sci. 0642
Peano, G. 0316
Pennsylvania Univ., Zool. Lab. 0796
Penzig, O. 0730
Period. sci. mat. (nat) etc. 0300; 108
Perthes' Geogr. Anstalt. *See* Justus Perthes' Geogr. Anstalt.
Perthshire Soc. Nat. Sci. 0647
Peterburg. *See* St. Peterburg.
Petermann, A. 0551
Petermanns geogr. Mitth. 0551; 110
Peters, C.A.F. 0329
Petit de la Saussaye, S. 0854
Petrograd. *See* St. Peterburg
Pettenkofer, M. 0884, 0925; 170
Pfalzbayr. Beitr. Gelehrsamk. 96
Pfeiffer, L. 0853
Pflüger, E.F.W. 0889; 167
Pharm. Centralbl. See *Chem. Centralbl.*
Pharm. journ. (and trans.) 0413; 105
Philadelphia
  Acad. Nat. Sci. 0598, 0616, 0831; 106, 111
  Amer. Philos. Soc. 029, 0106; 94, 121
  Entomol. Soc. 0818
Phillips, R. 064; 124
Philos. mag. 050; 103, 107, 124, 125, 130, 131, 146, 151
Philos. Soc. Adelaide 0156
Philos. Soc. Glasgow 0114
Philos. Soc. New South Wales 0195
Philos. Soc. Victoria 0161
Philos. Soc. Washington 0223
Philos. trans. 02, 094; 93, 97, 107, 115, 116, 119, 131
Physic. rev. 0388; 109, 146
Physic. Soc. Edinburgh 0766; 144, 162
Physic. Soc. London 0372, 0389; 146, 147

Physik. Arbeiten einträcht. Freunde 95
Physik. Ber. 0366, 0373
Physik. Gesellsch. Berlin 0362d, 0366, 0381; 109, 144, 145, 146
Physik. Med. Gesellsch. Würzburg 0143, 0236; 103, 144
Physik. Med. Soc. Erlangen 0191
Physik. Ökon. Gesellsch. Königsberg 0172
Physik. Tagebuch 95
Physik.-Tech. Reichanstalt 0380; 147
Physik. Verein Frankfurt 0104; 103
Physik. Zeitschr. 0390; 109, 146
Physiograph. Forening Christiania 080
Physiol. Gesellsch. Berlin 0905; 169
Physiol. Soc. 0893; 168
Phytologist 0696
Picard, J. 0351
Pictet, C. 048a
Pictet, J. 0129
Pictet, M.A. 048a; 126
Piria, R. 0365b
Pisa
  Soc. Toscana Sci. Nat. 0651
Pitagora 0322
Plant world 0749
Plantamour, P. 0401c
Plymouth
  Devonshire Assoc. Adv. Sci. 0177
Pochvovedenie 0539
Poggendorff, J.C. 0362c; 91, 144, 145
Polli, G. 0404c
Polsk. Towarz. Przyrod. 0227
Pop. astron. 0350; 108
Pop. sci. monthly 0217; 107
Posen
  Deutsch. Gesellsch. Kunst Wissensch. 0273
Poske, F. 0385
Potsdam
  Preussisch. Geodät. Instit. 0573
Powalky, K.R. 0356b
Prace mat. fiz. 0312
Prague. *See* Praha.
Praha
  Böhm. Gesellsch. Wissensch. 040, 0170; 94, 97
  Česká Akad. etc. 0262
  Jednota Česk. Mat. Fysik 0298
  Naturhist. Verein 0631
  Spolek. Chem. Česk. 0425
Praze. *See* Praha.
Preussisch. Akad. Wissensch. *See* Akad. Wissensch. Berlin.
Preussisch. Botan. Verein 0711
Preussisch. Geodät. Instit. 0573
Preussisch. Geolog. Landesanstalt 0505
Princeton Univ. 0258
Pringsheim, N. 0708
Proctor, R.A. 0235
Psyche 0821

Quart. journ. conchol. 0857
Quart. journ. lit., sci. arts 067; 124
Quart. journ. microscop. sci. 0675; 112, 158
Quart. journ. pure appl. math. 0281; 104, 108
Quart. journ. sci. 0183
Quekett Microscop. Club 0676; 112
Quesneville, G.A. 0410a,b
Quetelet, A. 0278

Rabenhorst, L. 0703
Racc. opusc. scient. filolog. 96
Ranvier, L. 0921
Rapp. ann. progrès (sci. phys.) chim. 0401c
Rassegna mineraria 0527
Real ... *See under the next word of the name.*
Record zool. literature 0771
Recueil mém. 022
Recueil pour les astronomes 95
Recueil trav. chim. Pays-Bas 0434
Recueil zool. suisse 0781
Regensburg
  Bayr. Botan. Gesellsch. 0685; 159
  Naturwissensch. Verein 0623
  Zool.-mineral. Verein 0623
Regia ... *See under the next word of the name.*
Reichel, J.S. 021
Reichert, C.B. 0872a; 167
Reil, J.C. 0869a; 166
Remsen, I. 0429
Rennes
  Soc. Sci. Méd. Ouest 0268

Repert. botan. systemat. 0697
Repert. Exp.-Physik [ed. Carl] 0368
Repert. Exp.-Physik [ed. Fechner] 0364a
Repert. mineral. kryst. Lit. 0501
Repert. Pharm. 0400; 105
Repert. Physik [ed. Dove] 0364c; 146
Repert. Physik [ed. Exner] 0368
Repert. physik. Technik etc. 0368
Reuss, J.D. 91, 98-100
Rev. Amer. chem. res. 0446
Rev. bryologique 0722
Rev. cours scientif. France etc. 0180
Rev. gén. botan. 0735; 161
Rev. gén. chim. 0453
Rev. gén. sci. etc. 0260; 107
Rev. géolog. 0485
Rev. mag. zool. 0764
Rev. math. 0316
Rev. math. spéc. 0315
Rev. prat. électr. 0386
Rev. rose 0180
Rev. scient. (France etc.) 0180
Rev. scient. ind. 0410a
Rev. semestr. publ. math. 0319
Rev. suisse zool. etc. 0781
Rev. zool. 0760
Revista ... See also *Rivista* ...
Revista chilena hist. nat. 0673
Revista mineraria 0476
Revista progresos ciencias etc. 0144
Rhodora 0751; 112
Riga
  Naturf. Verein 0123
Rijks ... *See under the next word of the name.*
Riunione Scienziati Italiani 0108
Rivista ... See also *Revista*.
Rivista geogr. ital. 0568
Rivista ital. paleont. 0528
Rivista ital. sci. nat. 0665
Rivista mat. 0316
Rivista mineral. crist. ital. 0514
Robin, C. 0882
Römer, J.J. 0681
Roma
  Accad. Lincei 0211; 107, 131
  Accad. Pontif. Nuovi Lincei 0132, 0211, 0252
  Colleg. Romano, Osserv. 0332, 0578; 104, 106
  Soc. Geogr. Ital. 0556
  Soc. Geolog. Ital. 0509; 110
  Soc. Sismolog. Ital. 0530
  Univ. 0890
Rose, V. 0396
Rosenthal, J. 0897
Rostock
  Naturf. Gesellsch. 0249
Rotterdam
  Bataafsch. Genootsch. Proef. Wijsbeg. 031; 94
  Nederland. Dierkund. Vereen. 0773
Rouen 97
Roux, W. 0916; 169
Royal ... *See under the next word of the name except in the following cases.*
Royal Institution 067, 0145; 124
Royal Soc. Canada 0237
Royal Soc. Edinburgh 044, 0120; 93, 97
Royal Soc. London 02, 094; 103, 107, 115, 131, 146, 153, 177. See also *Philos. trans.*
——— *Cat. sci. papers* 91, 101, 147, 155, 172, 174, 177
Royal Soc. New South Wales 0195
Royal Soc. South Australia 0156
Royal Soc. Tasmania 0138
Royal Soc. Van Diemen's Land 0138
Royal Soc. Victoria 0161
Rozier, F. 022, 030; 93, 98, 119, 123, 125, 148
Russ, K. 0839
Russ. arkhiv patol. klin. med. bakteriol. 0935; 114
Russ. Astron. Obshchest. 0347
Russ. Entomol. Obshchest. 0814
Russ. (Fiz.) Khim. Obshchest. 0422; 109, 146
Russ. Geogr. Obshchest. 0554, 0585
Russ. Mineral. Gesellsch. 0465

Sächsisch. Gesellsch. Wissensch. 0126, 0139
Sächsisch.-Thuringisch. Verein Vogelkunde etc. 0842
Sällskapets Fauna Flora Fennica 0650, 0653
Saikingaku zasshi 0934
Saint ... *See* St. ...
Salem. Mass.
  Essex Instit. 0615, 0642
  Peabody Acad. Sci. 0642
Sallo, D. de 01

Samml. chem. chem.-tech. Vorträge 0452
Samuelson, J. 0183
San Fernando
Observatorio 0357
San Francisco
Astron. Soc. Pacific 0343
California Acad. (Nat.) Sci. 0159
Sankt ... *See* St. ...
Scarpellini, E.F. 0134
Scelta opusc. interess. 032; 94, 98, 121
Schäfer, E.A. 0900a
Scherer, A.N. 0397
Schjellerup, H.C.F.C. 0286
Schlechtendal, D.F.L. von 0687, 0698
Schleiden, M.J. 076; 163
Schles. Gesellsch. Vaterländ. Cultur 061
Schlömilch, O. 0285
Schlözer's Briefwechsel 96
Schmiedeberg, O. 0891
Schneider, O. 0376
Schreiber, J.C.D. 0593
Schröter, J.C. 95
Schuberg, A. 0798
Schultze, M. 0883; 167
Schumacher, H.C. 0329
Schwalbe, G. 0877, 0912
Schwed. Akad. Wissensch. 014b
Schweigger, J.S.C. 0397d; 149
Schweiz. Blätter Ornithol. 0843
Schweiz. Entomol. Gesellsch. 0815
Schweiz. Gesellsch. Gesammt. Naturwissensch. 086
Schweiz. Ornithol. Verein 0843
Schweiz. Paläont. Gesellsch. 0497, 0517
Sci. Artium Instit. Bonon. 013; 93
Sci. gossip 0192; 107
Science 0244; 107, 130, 131
Scient. American 0122; 104, 107
Sclater, P.L. 0838
Scott. geogr. mag. 0564; 111
Scott. Geogr. Soc. 0564
Scott. Meteor. Soc. 0576; 106
Scott. naturalist 0647
Secchi, A. 0578
Seemann, B. 0705, 0713
Seemann. W. 0705
Selle, C.G. 95
Senckenberg. Naturf. Gesellsch. 0635, 0643
Service Carte Géolog. France 0519
Sidereal messenger 0339a
Siebold, C.T. von 0763; 163
Siena
Accad. Sci. 026; 94
Siezd Russ. Estest. (Vrachei) 0201
Silliman, B. (Snr) 069; 125
Silliman, B. (Jnr) 069; 125
Simonelli, V. 0528
Sirius 0335
Skandinav. Archiv Phsiol. 0907; 169
Skandinav. Naturf. Läkare 0107; 104
Skofitz, A. 0702
Smithsonian Instit. 0128, 0137, 0176
Soc. Agric. etc. Lyon 058
Soc. Amateurs Sci. Arts Lille 053
Soc. Anat. Paris 0871; 113, 166
Soc. Anat. Physiol. ... Bordeaux 0896
Soc. Arcueil 060; 122
Soc. Astron. France 0341; 109, 143
Soc. Belge Géolog. etc. 0513
Soc. Biol. Paris 0876; 107, 113, 166, 168, 169
Soc. Botan. France 0706; 106, 112, 160
Soc. Botan. Ital. 0699, 0741
Soc. Botan. Lyon 0717
Soc. Chem. Ind. 0433; 109, 152
Soc. Chim. Méd. 0405
Soc. Chim. Paris 0417; 105, 109, 151
Soc. Cuviérienne 0760
Soc. entomol. 0828
Soc. Entomol. Belge 0813
Soc. Entomol. France 0806, 0834; 106, 112
Soc. Entomol. Ital. 0820
Soc. Entomol. Ross. 0814
Soc. Etude Sci. Nat. Elbeuf 0664
Soc. Etude Sci. Nat. Nîmes 0649
Soc. Fauna Flora Fennica 0650, 0653
Soc. Franç. Minéral. 0503
Soc. Franç. Phys. 0371; 146
Soc. Fribourg. Sci. Nat. 0659
Soc. Geogr. Ital. 0556
Soc. Géogr. Paris 0540; 105, 111
Soc. Géolog. Belg. 0496
Soc. Géolog. France 0464, 0466; 105, 110, 153
Soc. Geolog. Ital. 0509; 110
Soc. Geolog. Milano 0636

Soc. Géolog. Nord Lille 0492
Soc. Géolog. Suisse 0517
Soc. Helvét. Sci. Nat. 086
Soc. Hist. Nat. Départ. Moselle 0619
Soc. Hist. Nat. Paris 0600; 158
Soc. Hist. Nat. Strasbourg 0607
Soc. Holland. Sci. 023, 0197
Soc. Imp. Naturalistes Moscou 0596, 0606; 106, 111
Soc. Ital. Fisica 146
Soc. Ital. dei Quaranta. *See* Soc. Ital. Sci.
Soc. Ital. Sci. 037; 94, 98, 121
Soc. Ital. Sci. Nat. 0636
Soc. Ligustica Sci. Nat. etc. 0261
Soc. Linn. Bordeaux 0603; 111
Soc. Linn. Calvados 0601
Soc. Linn. Lyon 0604; 106
Soc. Linn. Normandie 0601, 0637
Soc. Linn. Paris 157
Soc. Malacol. Belgique 0855
Soc. Math. Amsterdam 0319
Soc. Math. France 0297; 108
Soc. Météor. France 0575
Soc. Minéral. France 0503
Soc. Muséum Hist. Nat. Strasbourg 0607
Soc. Naturalisti Modena 0200
Soc. Naturalisti Napoli 0668
Soc. Pharm. Paris 0399b
Soc. Philomat. Paris 045; 104, 122
Soc. Phys. Hist. Nat. Genève 078
Soc. Physico-Med. Basil. 019
Soc. Public Analysts 0426
Soc. Quaranta. *See* Soc. Ital. Sci.
Soc. Reale Borbonica 070
Soc. Regia Sci. Berolin. 07
Soc. Regia Sci. Gotting. 020; 93
Soc. Roy. Botan. Belg. 0712
Soc. Sci. Arcachon 0867
Soc. Sci. Fennica 0109
Soc. Sci. Liège 0117
Soc. Sci. Lille 053
Soc. Sci. etc. Nancy 054, 0222
Soc. Sci. Upsal. 011; 94
Soc. Sci. Méd. Ouest Rennes 0268
Soc. Sci. Nat. Cherbourg 0148
Soc. Sci. Nat. Neuchâtel 0620
Soc. Sci. Nat. Saône-et-Loire 0656
Soc. Sci. Nat. Strasbourg 0607
Soc. Sci. Ouest Nantes 0670
Soc. Sci. Phys. Nat. Bordeaux 0160
Soc. Sismolog. Ital. 0530
Soc. Spettrosc. Ital. 0336; 109, 143
Soc. Toscana. Sci. Nat. 0651
Soc. Vaud. Sci. Nat. 0618; 111
Soc. Zool. France 0774, 0787; 163
Soc. Zool. Tokyo 0803
Sokolov, D.I. 0461
South African Museum 0804
South African Philos. Soc. 0231
Spengel, J.W. 0785
Sperimentale 0879
Spolek Chem. Česk. 0425
St. Gall
  Naturwissensch. Gesellsch. 0169
St. Louis, Mo.
  Acad. Sci. 0164
St. Peterburg
  Acad. Sci. 012, 055, 0101; 94, 97
  Akad. Nauk 0801
  Biol. Lab. 0920
  Geolog. Komiteta 0510, 0511
  Instit. Eksper. Med. 0913
  Mineral. Obshchest. 0465
  Obshchest. Estest. 0214; 107
  Russ. Astron. Obshchest. 0347
  Russ. Entomol. Obshchest. 0814
  Russ. (Fiz.) Khim. Obshchest. 0422; 109, 146
  Russ. Geogr. Obshchest. 0554, 0585
  Russ. Mineral. Gesellsch. 0465
  Soc. Entomol. Ross. 0814
State Cabinet Nat. Hist. Albany, N.Y. 0762
Stettin
  Entomol. Verein 0809, 0810
  Ornithol. Verein 0840
Stockholm
  Entomol. Förening 0822
  Geolog. Förening 0495
  Jernkontor 0458
  Kemistssamfundet 0438
  (Svensk.) Vetensk. Akad. 014, 074, 0118, 0216, 0401a, 0686; 94, 97, 126, 150, 159
Stone, O. 0301
Storia lett. Italia 96
Strasbourg/Strassburg
  Geolog. Landesanstalt 0515
  Soc. (Museum) Nat. Hist. 0607

Sturm, J.C. 116
Stuttgart
Verein Vaterländ. Naturkund. Württemberg 0125
Svanberg, L.F. 0401a
Svensk kem. tidskr. 0438
Svensk Vetensk. Akad. *See* Vetenskaps Akad. Stockholm.
Sydenham, T. 116
Sydney
Austral. Museum 0791
Linnean Soc. New South Wales 0652; 111
Royal Soc. New South Wales 0195
Sylvester, J.J. 0281, 0305
Symons, G.J. 0579

Tagsber. Fortschr. Natur- und Heilkunde 076
Tartu. *See* Dorpat.
Taschenbuch Ärzte Chem. Pharm. 0392c
Taschenbuch gesammt. Mineral. See *Jahrbuch Mineral. etc.*
Taschenbuch Scheidekünst. Apoth. 0392b
Taylor, R. 050, 0614; 124, 158
Telegraph. journ. etc. 0369
Termész. füzetek. 0654; 111
Termész. közlöny 0207
Terquem, O. 0283
Terrestr. magnetism 0588
Testut, L. 0900a
Teyler's Stichting 0196
Thomson, T. 064; 124
Thomson, W. 0281; 130
Thornton, W.H. 0301
Thüringisch Botan. Verein 0737
Tidskr. mat. fys. 0292
Tidsskr. math. 0286; 104
Tijdsch. entomol. 0811
Tilloch, A. 050; 124
Tissandier, G. 0220
Tokyo
Botan. Soc. 0734
Chem. Soc. 0432
Soc. Zool. 0803
Sugaku Butur. Kwai 0311
Univ. 0232
Torino
Accad. Sci. 038, 065, 0189; 94, 97
Univ. 0784; 169
Terrey Botan. Club 0716; 112
Tortolini, B. 0284
Toulouse
Acad. Sci. etc. 036, 087; 93, 121
Faculté des Sci. 0253
Towarz. Nauk. Krakowskie 0199
Trebra, F.W. von 95
Treub, M. 0724
Triesnecker, F. 0353
Tring Zool. Museum 0797
Trommsdorff, J.B. 0392c, 0395; 149
Tromsø Museum 0775
Tschermak, G. 0494
Turin. *See* Torino.
Turner, W. 0887
Tutt, J.W. 0830
Tychsen, C. 0286
Tyndall, J. 130

Uhlworm, O. 0725, 0930
Ungar. Akad. Wissensch. 0238
Unione Zool. Ital. 0792
United States. Bureau of Standards 147
United States. Coast (Geod.) Survey 0546; 156
United States. Dept of Agric. 0584, 0790, 0824; 164
United States. Fish Commission 0862
United States. National Museum 0780
Università Gregoriana. *See* Collegio Romano.
University ... *See under the name of the place.*
Unterrichtsbl. Math. Naturwissensch. 0321
Untersuch. Naturlehre Menschen Thiere 0878; 107
Uppsala
Soc. Sci. 011; 94
Univ. 0524
Usteri, P. 0681
Utrecht
Hoogeschool 0415
Univ. 0875

Van't Hoff, J.H. 0437
Varshava. Varsovie. *See* Warszawa.
Venezia
Istituto Veneto Sci. etc. 0112
Venice. *See* Venezia.

Verein Erdkunde etc. Darmstadt 0549
Verein Förder. Math. Naturwissensch. Unterrichts 0321
Verein Freunde Astron. etc. 0346
Verein Freunde Naturgesch. Mecklenburg 0625
Verein Naturkunde Cassel 0612
Verein Naturkunde Herzog. Nassau 0119
Verein Naturwissensch. Braunschweig 0233
Verein Vaterländ. Naturkunde Württemberg 0125
Verein Verbreit. Naturwissensch. Kenntn. Wien 0171
Vermont Geolog. Survey 0538
Vestnik opyt. fiz. element. mat. 0384
Vetenskaps Akad. Stockholm 014, 074, 0118, 0216, 0401a, 0686; 94, 97, 126, 150, 159
Victorian naturalist 0666; 111
Videnskabs Selskab. Christiania 0168
Vienna. *See* Wien.
Vierteljahrsschr. Fortschr. Gebiete Chem. Nahr. Genussmitt. 0436
Vines, S.H. 0732
Virchow, R. 0874, 0888; 167
Voigt, J.H. 035
Voit, C. 0884
Volta, A. 95, 121
Vriese, W.H. de 0701

Walch, J.E.I. 0593; 120
Walpers, W.G. 0697
Warsaw. *See* Warszawa.
Warszawa
  Varshav. Univ. 0212
Washington
  Acad. Sci. 0276
  Biol. Sci. 0895; 113, 169
  Entomol. Soc. 0825
  National Acad. Sci. 0182
  Philos. Soc. 0223
  Smithsonian Instit. 0128, 0137, 0176
  National Museum 0780
Watt, C. 0411
Watt, J. 0411
Weber, W. 0586
Werner, A.G. 95, 153
Wernerian Nat. Hist. Soc. 0597; 154, 157
West Riding Consol. Naturalists' Soc. 0632
Wetter 0583; 111
Whitaker, W. 0498
Whitman, C.O. 0906
Wiadomości mat. 0323
Wiedemann, G. 0362d; 144
Wiegmann, A.F.A. 0759
Wien
  Akad Wissensch. 0135, 0141, 0146, 0184, 0431, 0537; 103, 107, 131
  Geogr. Gesellsch. 0552, 0571; 105
  Geolog. Reichanstalt 0475, 0481; 105, 110
  Naturhist. Hofmuseum 0783
  Österreich. Gesellsch. Meteor. 0580
  Verein Verbreit. Naturwissensch. Kenntn. 0171
  Zool.-Botan. Gesellsch. (*or* Verein) 0633; 106, 111
Wiener Zeitschr. Physik etc. 0363
Wiesbaden
  Nassau. Verein Naturkunde 0119
Wikström, J.E. 0686
Willard, C.R. 0350
Wilson Ornithol. Chapter 0849
Winchell, W.H. 0516
Wisconsin Acad. Sci. etc. 0209
Wissensch. Meersuntersuch. 0589
Witzschel, B. 0285
Wöhler, F. 0401b, 0409; 150
Woodward, H. 0482b
Wszechświat 0240
Würzburg
  Physik. Med. Gesellsch. 0143, 0236; 103, 144
  Univ. 0886; 168
Würzburg. naturwissensch. Zeitschr. 0143

Yale College 125
Yearbook scient. learned soc. Gt Britain 0248
Yorkshire Geolog. (Polytech.) Soc. 0468
Young, T. 0354

Zach, F. von 0326, 0327; 142
Zeitschr. allg. Erdkunde 0548
Zeitschr. analyt. Chem. 0420; 105, 110
Zeitschr. angew. Chem. 0439; 109, 152
Zeitschr. angew. Mikroskopie 0680; 112
Zeitschr. anorg. Chem. 0440; 109
Zeitschr. Berg-, Hütten- und Salinwesen preussisch. Staate 0478
Zeitschr. Biol. 0884; 113, 168
Zeitschr. Chem. 0418; 105
Zeitschr. Elektrochem. 0444; 109
Zeitschr. Entomol. 0810
Zeitschr. (gesammt.) Naturwissensch. 0154; 103
Zeitschr. Gewässwerkunde 0570
Zeitschr. Hygiene (Infectionskrankh.) 0928; 114
Zeitschr. Instrumentenkunde 0380; 109, 146
Zeitschr. Kryst. Mineral. 0501; 110
Zeitschr. Malakozool. 0853
Zeitschr. math. naturwissensch. Unterricht 0295; 108
Zeitschr. Math. Physik. 0285; 104, 108
Zeitschr. Mineral. 0456a
Zeitschr. Morphol. Anthropol. 0912
Zeitschr. öffentl. Chem. 0447
Zeitschr. Oologie 0847
Zeitschr. Ornithol. etc. 0840
Zeitschr. Physik Math. 0363
Zeitschr. Physik verwandt. Wissensch. 0363
Zeitschr. physikal. Chem. 0437; 109
Zeitschr. physikal. chemisch Unterricht 0385; 109
Zeitschr. physiol. Chem. 0427; 109
Zeitschr. prakt. Geologie 0526; 110
Zeitschr. Untersuch. Nahr. Genussmitt. 0436; 109
Zeitschr. Vermessungswesen 0574
Zeitschr. wissensch. Mikroskopie 0679; 112
Zeitschr. wissensch. Zool. 0763; 112, 163
Zeitschr. wissensch. Botanik 163
Zemlevyedyenie 0567
Zeno, A. 08
Zentralbl. ... See *Centralbl.*
Zentralzeit. ... See *Centralzeit.*
Ziegler, E. 0901
Zool. Anzeiger. 0776; 112, 163, 164
Zool.-Botan. Gesellsch. (*or* Verein) Wien 0633; 106, 111
Zool. Garten 0769; 112
Zool. Gesellsch. Frankfurt 0769; 163-164
Zool. Jahrbuch 0785; 112
Zool.-Mineral. Verein Regensburg 0623
Zool. record 0771; 164
Zool. Soc. London 0756, 0757, 0771; 106, 112, 138, 162, 164
Zool. Station Neapel 0864; 165
Zool. Zentralbl. 0798
Zoologica 0788
Zoologist 0761; 106, 112, 162, 164
Zürich
  Concilium Biblio. 0533. 0802, 0915, 0918, 0919; 138, 164, 169, 176
  Naturf. Gesellsch. 0130